How Early
America
Sounded

How Early America Sounded

Richard Cullen Rath

Cornell University Press

ITHACA AND LONDON

First published 2003 by Cornell University Press

Printed in the United States of America

Library of Congress Cataloging-in-Publication Data
Rath, Richard Cullen.
  How early America sounded / Richard Cullen Rath.
      p. cm.
    Includes bibliographical references and index.
    ISBN 0-8014-4126-9 (alk. paper)
    1. United States—Social life and customs—To 1775.   2. United States—Social conditions—To 1865.   3. Sound—Social aspects—United States—History.   4. Voice—Social aspects—United States—History.
5. Hearing—Social aspects—United States—History.   6. Listening—Social aspects—United States—History.   I. Title.
E162 .R38 2003
973.2—dc22   2003019789

Cornell University Press strives to use environmentally responsible suppliers and materials to the fullest extent possible in the publishing of its books. Such materials include vegetable-based, low-VOC inks and acid-free papers that are recycled, totally chlorine-free, or partly composed of nonwood fibers. For further information, visit our website at www.cornellpress.cornell.edu.

Cloth printing   10 9 8 7 6 5 4 3 2 1

To Monisha

Contents

Preface

Sound was more important to early Americans than it is to you. Their world was in many respects a quieter one. There were no radios, no cars, not even the hum of a refrigerator, a fluorescent light, or a computer. But there were other sounds that would be unfamiliar to you: the handbell and calls of the town crier, women batting their laundry, or horses' hooves and ironclad wheels on cobblestones. You, on the other hand, might turn on a radio or television to drown out the roar of a nearby highway. The noises of appliances and other electronic devices and equipment escape your hearing unless they are brought to your attention. The world of early America was also darker than yours. When the sun went down, only dim lamps and fires remained. Your neighborhood might be lit up as bright as day every night. Technology, bright and loud, has attenuated your auditory and accentuated your visual perceptions of the world about you. Throughout the seventeenth century, however, sound had a more immediate power that we no longer associate with it: it was a tangible force laden with intent rather than your harmless whiff of disturbed air.

Perhaps you find my addressing you directly here as peculiar, perhaps even intrusive. I might have chosen instead to introduce this book by writing that "sound was more important to early Americans than it is to present-day Americans," allowing you a comfortable distance from the text, giving it the appearance of a thing to be consumed, evaluated, and judged. In contrast, sixteenth- and seventeenth-century chroniclers were likely to narrate a book as I have, saying, "You have heard" rather than "We have seen." For a reader to have "heard" implies sound: before, and perhaps during, the reading of such a book it would have been assumed that you were as likely to listen to ideas as to see them. The use of "you" interrupts the modern objective gaze that we have come to take for granted. That gaze is steeped in the culture of print and reading, where the eye is the primary organ of perception.

Here and there in this book I have indulged in the conceit of talking about a somewhat homogenous modern or postmodern "us" that perceives the world differently than the subjects of this book. The "we" in question is highly literate and visually oriented. In using this construct, I in no way mean to infer that you have grown deaf in some irretrievable fashion. In fact, I rely on you to hear the page as well as see it. This act of synesthesia is only possible to the extent that your auditory imagination is in working order, so it is not a deafness that I propose, but a loss of some measure of hearing's importance, and the setting aside of older ways of hearing that has taken place over many generations. I ask you to pick these ways up from their corner, as it were, dust them off a bit, and consider them anew, perhaps like another synesthetic medium, an old vinyl record.

For its part, early modern reading was a different and much more variegated species than it is today, centered on hearing the page as much as seeing it. Just how different it was is only partially demonstrated when you are confronted with elements of print style that were perfectly acceptable then. Texts used *italic,* 𝔟𝔩𝔞𝔠𝔨𝔩𝔢𝔱𝔱𝔢𝔯, and **bold** typefaces to indicate verbal emphases in seventeenth-century print, giving sonic cues we now find strange. Reading itself was often performed out loud, so that even the illiterate could benefit. Learning to read was an intensive process of sounding out letters, and for most non-elite readers it may have stayed that way. Silent reading was perhaps the exception rather than the norm.

Maybe you have let my direct address pass, this being the book's preface. Over time, direct address in books was relegated to prefaces, which were meant to frame the book for the readers. Then even this vestige disappeared, even though frames of this sort are generally conservative. There is more at stake here than the preface.

The subject of this book—how people heard their worlds in early America—requires an ability or at least a willingness on your part to be repeatedly decentered. It requires you to think about the process of taking in knowledge through the eyes from a black-and-white, two-dimensional page, and to think of that act as a socially learned rather than natural way of coming to know things, one that has had profound effects on how you perceive worlds beyond the printed page. This denaturalization, this historical situating of your own reading and its cognitive effects, is a necessary starting point for coming to grips with a sometimes alien world of powerful sounds. Thus the task with which I charge you is to be conscious of the visually centered ways you silently read, to set your habits aside here and there in order to hear more clearly how early America sounded—not to us, but to its denizens.

Introduction

In 1666, Samuel Arnold, the minister of Marshfield in Plymouth Colony, wrote to Increase Mather describing a panicky household as it suffered through a fatal summer storm:

> The woman of the hous calling earnestly to shut the dore which was done, instantly a terrible clap of *thunder* fell upon the house & rent the chimney & split the doore in many places & struck most of the persons if not all.

Writing about this document in 1850, the New England historian Nathaniel Shurtleff attributed the damage to a stroke of *lightning,* silently translating the audible clap into a visual stroke. He is consistent in describing the damages as "caused by lightning," as "the effects of lightning," and as "happened, through the agency of lightning," although his sources are nearly equally consistent in attributing the damage to thunder.[1]

The texts that Shurtleff translated were not alone in attributing a tangible power to the sound of thunder. In 1891, the New England historian Sidney Perley noted that in Cotton Mather's day, "it was generally supposed that thunder and not lightning caused the damage." After Perley, the distinction seems to have slipped from historical consciousness altogether. Writing a decade later, Reuben Gold Thwaites silently and probably unconsciously substituted "lightnings" for the "thunders" (*tonnere*) reported in the original French of the *Jesuit Relations,* yet he meticulously translated the plural, which had an important meaning in the Native American soundscape being described. And today, lightning is often substituted for the thunder of seventeenth-century sources, probably in unconscious deference to modern readers.[2] In fact it was this dissonance between interpretation (lightning) and source (thunder) that first suggested to me that a shift in perception may have taken place.

These silent print translations pose two related problems. The first is how to come to terms with the sensory world as it was before it needed to be translated. Sound mattered to early Americans in ways that it no longer does, particularly in the years between 1607 and 1720 or so. The task here is to tease out these older soundways in an effort to recover a portion of the seventeenth-century sensorium, one muffled by time, documentation, and the literate, highly visual mindsets of scholars. I do so by studying soundways: the paths, trajectories, transformations, mediations, practices, and techniques—in short, the ways—that people employ to interpret and express their attitudes and beliefs about sound. I am not so much concerned with the underlying beliefs, historically inaccessible as they often are, or the concrete expressions themselves so much as the ways between them. This approach places the book squarely in the domain of cultural history. As such, it is a contribution to the history of the senses, a field long called for, but only recently undertaken with any seriousness.[3]

The second problem is how to understand the nature, causes, and timing of the shift from sound toward vision. A prerequisite for such an understanding would be a solid foundation in a preexisting world with a heightened or very different soundscape, a world strange enough to us as to require explanation. Building that foundation by way of recovering the sounded worlds of early Americans is a sufficiently daunting task, one that will require a flexibility and willingness to think differently on the part of the reader. For this reason I have set aside anything more than a brief explanation of the shift itself, leaving that problem for another time. Interesting as it is in its own right, it deserves a separate treatment. The focus here will be to construct a base upon which such a transformation could take place.

One account of why such a shift took place does need to be considered, though. Marshall McLuhan, the media guru *par excellence* of the 1960s—whose theories have undergone somewhat of a renaissance as the digiterati have found new relevance in them—asserted that print and literacy gave humans "an eye for an ear," shifting the proportions of the sensorium.[4] Humans without literacy are part of an ear-based oral culture. These oral cultures have not undergone, via print and literacy, the transformation to the modern, visual world, with its abstract, objectified knowledge and reflexive thought. As such, oral cultures are in a state of nature, and people in them are constrained to particularly oral ways of thinking, so the theories go.[5] A culture in a state of nature is unchanging and thus ahistorical. Anthropologists and historians have relied on the theory of orality to say much about societies where print and literacy were rarer, such as Native American peoples, African American communities, non-elites, and women.

To the extent that they were illiterate they are deemed pre-modern and thus oral.[6] Literacy and print in turn have been associated with the development of western civilization, in which case oral cultures stand in for the primitive and the "savage." Illiteracy becomes grounds for colonization.[7]

This literacy hypothesis hinges on the assumption of an older, ear-based way of life. Without that, there is no shift in perception, and without that, no literate/oral divide. Yet orality is not established empirically in these theories; it is established by inference. It is what literacy is not, and serves as a foil. The evidence for the transformation of ear-based oral culture to visual print culture is thus circular. Orality is itself the product of literate minds. So-called oral cultures would have no need for the term. Oral culture comprises that part of spoken language that could have been printed or written but was spoken instead. Any part of the audible world outside that which is reducible to print and writing gets missed. And where print and writing create objects—namely, texts—oral expressions are ephemeral, disappearing even as they are made. As such, so the theory goes, orality cannot be documented, and to write or print it disqualifies it as oral culture.

But not all sounds can be reduced to print. Thunder and other natural sounds are a fine example. Presumably they sound more or less the same now as they did in early America, so the problem of sound being evanescent is moot. What has shifted is how they are heard. Thunder and other natural sounds did things then that we no longer allow them to do.

Throwing a wrench in the literacy hypothesis, natural sounds had these powers even in the most literate culture in the world at the time, that of the New England Puritans. There *was* a shift in the ways of interpreting thunder and lightning, but it came long after print and literacy. Perhaps these natural sounds are an exception, so let us consider some other sounds.

Instrumental sounds—that is, sounds made by humans with nonhuman devices—were often used to constitute community. Bells rang people in; militias could drum them out. While some of the sounds have been lost, others in the form of old bells or acoustical spaces can still be heard today. A bell reproduces an old sound as surely as a document reproduces an old thought. The inside of a meetinghouse reverberates today just as it was designed to centuries ago.

For example, the Puritan John Gyles, held captive as a boy for seven years by Micmac Indians and the French in Eastern Canada during the 1690s, wrote a short book about his experiences. In one part, he described happening upon what sounded to him like "a Woman washing her Linnen with a batting staff." He was in the deep forest, though. As he investigated closer, he found that the sound came from land turtles "propogating." He had, he claimed, heard them from half a mile away. Presumably the turtles would sound the same today. Gyles was writing for an audience that he assumed knew the sound of batting staffs on laundry, a sound no longer common to life in the twenty-first century. The turtles let us listen in not only on their amorous adventures, but on a sound culled from everyday life, one that marked the hearer as being within half a mile or so of a familiar community.

By considering instrumental sounds like bells and batting staffs we reintroduce two of the features of orality, intent and human construction, without yet getting caught in its circular logic. Like thunder, instrumental sounds held meanings in early America that we have lost. But just as the sounds themselves can sometimes be recovered, so too can the ways of interpreting them. We can still get closer to orality without giving in to its flaws, however.

Early Americans were captivated by sounds of voice as they existed alongside or beyond the realm of language. For example, a Pamunkey Indian way of healing, at least as the colonist John Smith described it in the early seventeenth century, was for "a man with a Rattle and extreame howling, showting, [and] singing" to dance and suck out the blood and phlegm from the infected place.[8] Smith was not concerned with oral culture here, although he was describing voice. He did not know the language, so he could not reduce it to writing. What he did write about was how that language of healing sounded. By saying that the Indian howled, he placed him outside the realm of the civil, in the "howling wilderness." How one's voice sounded was a good indicator of where he or she was located in early America: Ranters lived on the edge, murmurers and grumblers were a threat from within, and those whose voices escaped language altogether were wild or savage, unless they were groaning toward God, which placed them between the earthly and the heavenly.

Women were particularly susceptible to the vocal nonverbal sounds of others. Take the case of Winifred Holman. In 1659, one of her neighbors accused her of being a witch. The charge was based on the sounds and quietness of an infant who could not even talk yet. Holman's neighbor had a small daughter who "was taken with a strange raving and marvellous unquiet night and day" after Holman had visited her. The neighbor further "observed that when Mrs. Holman and her daughter were gone abroad that she [the allegedly bewitched infant] was pretty quiet." The child's inarticulate sounds were enough to get Holman brought to trial for witchcraft. She was acquitted on the testimony of her neighbors that she was "a diligent hearer and attender to the word of God," but the fact of the accusation itself shows how vulnerable women were to the sounded world.[9]

These vocable sounds—lying outside the domain of orality or that which can be reduced to print and writing—bring us another step closer to a historical understanding of early American soundscapes. Vocables located people in terms of civil society and the heavens—something such sounds no longer have the power to do. These sounds are documented. Even without hearing the actual tones, we can sometimes recover their meanings.

Now when we factor the articulate aspects of speech back in, they can be embedded in a rich historical context rather than existing in an ahistorical state of nature. Soundways and soundscapes replace crude notions of orality. For the most part I have left speech out of the book. Numerous scholars have written ably about it in recent years,[10] and from them we have learned that speech and orality were important in early America. My focus on nearly all but the linguistically articulate parts of the soundscape tells us *why* they were so important. A few examples are constructive nonetheless. Consider the case of Anne Hutchinson. Between 1636 and 1638, she was embroiled in a conflict that has come to be called the Antinomian Controversy. It concerned her preaching on Sundays outside the regular meeting, where she began to draw a significant number of followers. Boston's elites found such preaching, especially coming from a woman, to be profoundly disquieting. Governor John Winthrop argued that her voice must be stopped in order to "hold fast the sound form of words." It was the "constant language of scripture," he asserted, which provided truth and "soundness of phrase." This female threat to sonic order, thought Boston's male leaders, was liable to start an uprising like the Anabaptist insurrection that had taken over Münster a century earlier. Hutchinson's "voluble tongue" was more of a threat than the arms of all her followers. They were told that if they were to renounce their allegiance to her, they would be allowed to keep their guns. Hutchinson herself was banished.[11] Her case demonstrates not only the importance of speech in the first generation of New England society, but also how it was gendered.

In European American society, elite and commoner were knit together in a world of powerful sounds and speech acts. Even when they were still in England, the Puritans were concerned with legislating speech, linking social order to spiritual grace. In 1629, Matthew Cradocke, the London-based governor of the "Plantation in the Massachusetts Bay," wrote to John Endecott, one of the earliest settlers, imploring "yow who are in authoritie" to "make some good lawes for the punishing of swearers." Cradocke feared that many of the servants and non-elite immigrants were addicted to swearing, which needed to be reformed "if ever you expect comfort or a blessing from God upon or plantaccon."[12] Elite control of spoken utterances was thus bound to both the spiritual and social well-being of the community.

In his perennially reprinted book of manners, Richard Allestree said that the first duty of seventeenth-century servants was to obey the sound of their master's voice. They were supposed to submit to the master's rebukes "not answering again," even when they were "undeserved reproofs." They were to always have "given the Master the Hearing" that an elite supposedly deserved by dint of his class position. Allestree directed masters to moderate their use of this power. When they had to rebuke a servant, their voices should remain cool, reasoned, and calm. The tone of their commands was to be temperate, never "heated." The servant's obedience, continued Allestree, "must not be a grumbling and unwilling one." Grumbling was in the tone of the voice as much as the content. It was done at the lower ends of the social hierarchy, and it was always directed upward. Inferiors grumbled discontentedly at their superiors rather than silently showing deference.[13]

The fact that such a prescriptive book existed and sold well indicates a need for it. Prescriptions are remedies, and remedies are not needed if nothing is wrong. The lower sorts had to be told not to grumble—or whisper or murmur—which indicated that they did these things sometimes. This points toward class relations that were not entirely in conflict. But neither was there a rank order society in which social hierarchies are accepted norms. A sonic conception of this sort of class relations is that there was dissonance, but it manifested itself as tension rather than antagonism. From the masters' perspective, grumbling was dissonant, while from the servants' perspective it was the sound of the masters that jarred. Both were part of the same composition, though. For that reason I have looked for class relationships more than class positions.

There were regional differences between North and South, and to some extent, within those regions as well. Seventeenth-century Puritans bordered on being obsessed with controlling their soundscapes. Quakers in the middle colonies developed soundways based more on reciprocity than any other region. In the Chesapeake, the first generation of settlers were highly aware of their soundscapes. Thunder, bells, gunshot, howls, and speech all played key roles in the comprehension and construction of a new world. Southern elites seem to have shifted toward vision earlier than northerners, but non-elites held on to the power of sound much later, well into the eighteenth century. Yet like their northern counterparts, southern elites needed to stay connected to the rest of their society, if only to try and lead it. Eighteenth-century southern courts and plantations played out an audible drama—courts "heard" cases—that served this function well.[14]

Europeans were not the only early Americans whose soundways can be tracked. A few words need to be said here about the methods and scope I have used for non-European soundscapes. The sources are rarer and harder to read because of biases, but Native American and African American soundscapes turn out to be distinctive and historically situated much more than a narrow focus on orality would allow. While there are changes over time, especially in treaties, there are discernable continuities in Native American soundways well into the eighteenth century. For this reason, I have included eighteenth-century documents as well as those from the seventeenth century to bolster my arguments concerning Native Americans. The documentation for African American soundways in the seventeenth century is thin. "Upstreaming"—that is, inferring from later sources—does not make sense because until the last decades of the seventeenth century, slavery remained a marginal, albeit growing, institution in the North American labor force. African American life, North or South, changed in the eighteenth century, as southern plantation economies transformed the demographics and experience of slavery radically. During the seventeenth century, most enslaved Africans in the northern hemisphere were located in the Caribbean, with Jamaica leading the way at this time. From an African-centered perspective the focus on Jamaica makes sense because that is where the Africans were. In addition, enslaved Africans who did land in British North America had often done stints in the Caribbean first. I have drawn on eighteenth-century evidence for African American soundways in North America, but I have also relied heavily on a few rich Caribbean and African sources to piece together a provisional African American soundscape for the late seventeenth century.

The Iroquoian and Eastern Woodlands Indians that I have studied generally associated sound with identity. Thunder marked the presence of thunderers, great spirits that sometimes took the form of birds and played ball and hunted huge snakes (the lightning) from above the clouds. Songs constituted identity and to have one's song broken, by torture or punishment, was tantamount to losing one's self. The meaning of the words did not matter as much as saying them. Words that were all right for Christians could kill Indians.

For example, a well-known Native American shaman from north of Quebec reported his dreams to the missionary Paul Ragueneau—dreams that convinced the Indian to reject Christianity: "I saw on several occasions last winter, the Manitou who governs the birds, the fishes, and the animals." As long as he sang the song that came to him in sweat lodges, the spirit promised him plenty. "In fact, . . . so long as I sang and beat my drum, my traps for Bears, for Beavers, and for other animals, never failed me." Other Indians had died of hunger and disease "because they amused themselves with certain words or certain prayers that were taught them" by the missionaries. The song he sang in sweat lodges and before hunting was his own, a marker of who he was. By respecting who he was and the spirits of the animals he hunted, he claimed, his life remained in balance. Uttering the wrong sounds—the foreign words of prayers and the religion of the missionaries—threw other Indians out of balance by attaching sounds to them that were not integral to who they were, displacing sounds that were. The results were disastrous.[15]

In 1688, on a plantation in Jamaica, a single enslaved African sang the words "*Hoba Ognion*" repeatedly in time to homemade string and percussion instruments. Each time he repeated the words, the rest of the company of bound Africans would clap their hands and sing "*Alla, Alla*" in response. This pattern was a key part of communal consciousness in African American social settings. The call and the response bound the singer together with the rest of the people even though rifts of ethnicity and language conspired to separate them. On most occasions there was no audience, only participants.[16]

The song's words carried many meanings, not all of them comprehensible to everyone in the group. *Hoba* may have meant "man," "child," "someone thought to possess other-worldly skills and contacts," "a secondary deity," or "the local community with which one identifies." *Ognion* may have meant "to copulate," "a form of witchcraft that calls on ancestor spirits or foreign deities," or "to wander." In modern Anang (West African) religions, *obio ekpo* refers to the community of souls of the properly dead, while *ekpo onyon* means a homeless, wandering soul, or ghost. By inference, *obio onyon* would be a homeless wandering community—living ghosts—a compelling, if speculative, subject for slave song. Perhaps the only deities left to be invoked by Jamaican slaves were foreign gods and wandering ancestral spirits. *Ognion* may have been a reference to the rootlessness of the new community, but it may also have been a conjuring of spirits who might have wandered as far as these Africans had. The refrain "*Alla, Alla*" may have been an entreaty to the Muslim god. Islam had drawn converts in West Africa before the seventeenth century, but had not reached Central Africa. A number of Arabic/Muslim words remain in Jamaican Maroon vocabulary today.[17]

The composite of these terms is resonant both with the situation in which Africans found themselves in Jamaica and with the various cultures from which they came. *Hoba Ognion* may have referred to the ancestral, communal, spirit-identified self—"we," not "I"— as well as a sense of homelessness, a loss of location and community. The refrain may have been an invocation of a foreign god, who was nonetheless not the god of the slave owner. None of these meanings can be pinned down with any certainty. The song seems to be an agglomeration of possible meanings from loosely related or unrelated languages and cultures. The words carried no single meaning to the enslaved, who spoke several different languages and came from different places. Perhaps they were cobbled together on the spot. What was important in this setting was the sound, the coming together of all these disparate strands not in linguistic comprehension but in a mix of harmonies and dissonances carrying meanings and unmeanings, some linguistic, some not, yet still producing a coherent whole.

This book's organization follows the pattern mapped out at the beginning. The first chapter explicates the natural soundscapes of early America. This is the realm of nonhuman sounds: thunder, waterfalls, wind, and earthquakes. The second chapter listens in on how Early Americans used instrumental sounds to build and maintain their communities. Chapter Three explores another type of instrument, the acoustic design of meetinghouses. The fourth chapter factors in voice but not language to attend to ranting, groaning, murmuring, and so forth. Finally, the fifth chapter focuses on the audible world that Native Americans constructed for themselves and in response to the incursions of European Americans. Songs of identity, treaties, wampum, and othering are its subjects.

"The past," wrote novelist L. P. Hartley, "is a foreign country. They do things differently there."[18] I have proposed not just another country but other worlds. At the beginning of each chapter I have included some personal experience that helped me comprehend these worlds, sometimes on a visceral level. (Otherwise I have stayed out of the narrative.) My hope is that by attending to soundways I have been able to open up parts of these worlds, not to get a glimpse of them but to listen in. These were worlds much more alive with sound than our own, worlds not yet disenchanted, worlds perhaps even chanted into being.

CHAPTER ONE

~~~~~~

"Those Thunders, Those Roarings":
The Natural Soundscape

When ratling Thunder ran along the Clouds;
Did not the Saylers poore and Masters proud
A terror feele as strucke with feare of God?
Did not their trembling joynts then dread his rod?
Least for foule deeds and black mouth'd blasphemies,
The rufull time be come that vengeance cryes.

> —Translation of Lucretius, quoted in
> John Smith's *Generall History*

A tempestuous noise of thunder and lightning heard. Enter a
shipmaster and a botswain.

> —First words of Shakespeare's *The Tempest*

On a summer night, when I was seventeen years old, I got caught outside in a violent thunderstorm. Twice thunder cracked from a strike less than twenty-five feet away. It would light up the ground like bright daylight for a half second as I heard and viscerally felt the deafening split like a powerful force running through my body. It seems easy to imagine that had I been any closer it would have literally thrown me back. Perhaps a dozen times more, lightning struck very close by. The hair on my arms stood on end from the electricity in the air. This terrifying experience overwhelmed my sight, hearing, and touch all at once. The audible, tactile, and visual aspects were inseparable. For me, Samuel Arnold's 1666 description of the thunderstorm he had experienced captured the immediacy of the lived moment much better than Nathaniel Shurtleff's 1850 translation of that account for modern readers like himself—and us. As we consider the early American sounds that humans did not—and could not—produce, I would

*like readers to think of those sounds as the powerful experiences they were rather than
dry words on a page.*

S eventeenth-century North Americans listened carefully to sounds
that we would now consider "natural"—that is, unintentional sounds,
not made by humans. Earthquakes, wind, water, and especially thunder: each was powerful in its own ways. Today we attribute such sounds to
friction, weather fronts, and electrostatic discharges: in short, to the un-
thinking and immutable laws of physics. They are reactions rather than actions. In contrast, many seventeenth-century Europeans considered these
sounds to have intelligent sources with intent and power, even if such
sources were invisible. In fact, they attributed all sounds to some intentional being. In addition, they granted sound a power in which we no
longer believe. Sounds did things in the world. They moved people about,
struck them, and in the case of thunder, actually killed.

The inclination to write off such aural beliefs as just so much ignorance
and superstition needs to be checked. An attractive explanation along
these lines is that the colonists were simply mistaken in thinking that thunder was dangerous, and that advances in scientific knowledge have shown
lightning to be the true culprit. But that is the wrong answer. Imagine for
a moment a discussion between two deer, arguing about whether their
stricken companion had been felled by the blast of the horn or the light
from the high beams, never even thinking of the now-passed truck as a
cause. Thunder and lightning are merely the sonic and visual dimensions
of what is currently known to inflict the damage: electricity. That it now
seems more logical, scholarly, and correct to write of the efficacy of lightning than that of thunder is a reflection of a historical shift in our sensory
perceptions rather than a simple indicator of forward progress.

At the outset of the seventeenth century, the English accounted for it
in two general ways, one drawing on mechanical, the other on spiritual
causes. These two interpretations were not mutually exclusive. The mechanical explanations did not depend on the visual criterion of observability demanded today: sounds were palpable. Spiritual explanations took
it as an axiom that all sounds had willful agents at their source. Within both
the material and spiritual interpretations there were competing explanations as well.

Simon Harward, a resident of Banstead, a small rural town south of London, summarized a typical set of these beliefs in his 1607 pamphlet, *A Discourse of Lightnings*. He was writing to reassure the people of a neighboring
town after their church had been struck during a storm. Because he was
outlining various beliefs about thunder and lightning, some of which with

he agreed and others he did not, his pamphlet is a fair indicator of the gamut of English and continental beliefs at the outset of English colonization projects.[1]

In early-seventeenth-century mechanistic explanations of thunder, sounds physically acted on one another, and on material objects, too. Thunder behaved as a tangible force. The peals of a bell or the report of a gun might be used to disperse it. Cracks of thunder were thought to do physical damage, including killing people and destroying buildings. In the mechanistic—or as he called it, "philosophicall"—part of Harward's account, thunder occurred when the planets somehow lifted watery vapor along with "fiery spirits and exhalations" from the earth into the upper atmosphere, described as a very cold place. There, the vapor "is thickned [*sic*] into a cloud, and the exhalation (which was drawne up with it) is shut within the cloud." The hot noisome air that had been drawn up, unable to find passage out of the now-solidified and cold exterior of the cloud, had to force its way, according to Harward. If the "sides" of the cloud were thick, and the hot air plentiful and dry, then the escape would be marked by thunder with lightning. But "if the clowd be thin, and the exhalation also rare and thin, then there is lightning without thunder."[2] Lightning without its accompanying thunder was weak. Sound was at the source of its power.

The mechanistic explanations had distinctly gendered overtones. William Strachey's account of the *Sea Venture*'s exploits on its trip to Virginia (discussed below) was originally a private letter sent to an unnamed English gentlewoman, possibly the wife of a well-placed Virginia Company patron. On the whole, this private letter (published in 1625) was much more steeped in sonic description than his other Virginia history, which he wrote for an audience whose gender was unmarked. It was also more sonic than the only other publication about the *Sea Venture*'s trip, which, like Strachey's Virginia history, was written for a general audience. In the letter, Strachey discussed the other-worldly aspects of the storm's thundering and roaring as well as the practical effects of the din, but he never raised the "mechanical" issues. He was highly technical in his Latin description of the diabolical causes for the excess of thunder in the Americas, so it was not a question of his speaking down to his addressee—he was in fact addressing a well-educated woman of higher rank than himself.[3] Mechanistic accounts of thunder and the power of sound were the product of secular learning available only to elite males in the early seventeenth century.

Harward, who had to establish the legitimacy of his authorial voice— quite literally, his authority—highlighted the "philosophicall" explanation of thunder. Yet he was trying to reach and reassure, rather than distance himself from, a more general readership. As if to offset his readers' suspicion that he might be too bookish or elite, Harward concluded his material explanation by likening thunder to familiar sounds that both he and his readers knew well. His homely analogies show how the early modern English soundscape is foreign to ours, yet still familiar and in many ways recognizable. Thunder cracked aloud like "a Chestnut in rosting among cinders," or like "a bladder filled with air, being violently broken," or as "When green wood is burned, [and] the spirits burst out with some little crack," or like the much louder sound of "gunpowder issuing out of ordinance." Of course, " the clowdes then which far exceed the greatnesse of mountaines must needs give out a more forcible roaring."[4] Size was not the only reason that thunder was louder than a gunshot or a burst balloon, though.

While mechanical understandings of sound explained thunder's immediate workings, it was to Harward's reckoning ultimately a spiritual force, the willfully deployed voice of God. "There is added," he confided in his spiritual account, "a more principall operation" than the mechanical forces, namely "the handie worke of God, whereupon thunder in the scriptures is called *the thunder of God*," and "*the voice of thy* [i.e., God's] *thunder.*" He notes further that "*the Lord thundred* [sic] *out of heaven and the most highest gave out his voice, hailstones, and coales of fire.*" Ending his explanation of this first cause of thunder and lightning, Harward quoted the divine admonishment of Job, in which God demanded to know who it was that divided "*the way for the lightenings of the thunder.*" Harward's biblical quotation categorized lightning as a property of thunder, the opposite of today's conceptions of the phenomenon.[5]

Although Harward had definite opinions about the true interpretations of thunder and lightning, he presented a range of views with which he disagreed, too. He acknowledged, but discounted, the idea that thunder and lightning were caused by devils. He also dismissed the idea that the damage from the storm indicated God's particular displeasure with the town it had struck. While God was the first cause, the storm was not, as some feared, an act of divine retribution. Harward thought it was rather a test of faith, much like those Job endured, for although the church had been burnt, no people were hurt and a general conflagration never broke out.[6]

Seventeenth-century Englanders treated thunder as a speech act on the part of God or perhaps demons. Speech acts are utterances that do something in the world. Behind every speech act is an actual or implied "I" that performs the act. In the seventeenth-century Christian natural soundscape, that "I" was most often the Christian God, who spoke worlds into being and took lives with his voice of thunder. These performative speech acts had what the philosopher of language John Searle calls "perlocutionary force," which meant they had an actual effect on their objects. The scope of such utterances, and the types of entities they could operate upon, have narrowed greatly since the seventeenth century. Then, natural sounds—which emanated from the speech acts of the invisible world—could break buildings, judge, and kill.[7] Sometimes they could shape the fate of a whole colony.

A Sea Wreck

In 1609, two years after the founding of Jamestown, a convoy of seven large ships and two pinnaces (small scouting ships) set out from England with sorely needed supplies and more settlers for the struggling colony. Aboard the flagship *Sea Venture* were George Somers, the leader of the expedition, and Thomas Gates, the new governor. Gates carried instructions from the Virginia Company for bringing order to the infant colony, which was beleaguered by dissent and illness from within and from without by its ambivalent relations to the Powhatan people upon whose territory the colonists encroached—and upon whose goodwill they depended for food in the absence of English supplies.[8]

Just a few days shy of Virginia, the convoy fell into "a taile of the West Indian Horacano" near Bermuda, where the storms were known to "rather thunder than blow." One gentleman aboard the *Sea Venture,* William Strachey, described the storm so compellingly that it may have later inspired Shakespeare's *The Tempest.* The sounds of the hurricane played a critical role, not only in the immediate outcomes, but in the government of Jamestown as well. At the storm's onset Strachey remarked that "the wind singing and whistling most unusually" had caused the *Sea Venture* "to cast off our pinnace," which was in tow. One ship was thus lost even before the hurricane had descended in earnest. "A dreadful storm and hideous" immediately ensued, "swelling and roaring as if it were by fits." Immediately, the sound of the storm made communications onboard impossible. The "clamours" of "women and passengers not used to such hurly and discomfort" and the prayers and shouts of the more seasoned crew were all "drowned in the winds and the winds in thunder." There was "nothing heard that could give comfort."[9]

In the seventeenth century, the possession of reason depended on an attenuated and governable soundscape, but the sounds of the storm "overmastered the senses of all," according to Strachey. "The ears lay so sensible to the terrible cries and murmurs of the winds" that even the most seasoned sailors were terrified and shaken. The thunder, in turn, drowned out the winds and the roaring sea. This constant din, lamented Strachey, "worketh upon the whole frame of the body," laying a sickness upon it "so insufferable" that it "gives not the mind any free and quiet time to use her judgment and empire."[10]

The roaring sea, howling wind, and constant thunder also drowned out ship-to-ship communications, with dire consequences. Normally, during times of calmer weather, convoys communicated via flags, drums, trumpets, and shouting. At night, flashing lanterns replaced the flags. But in poor visibility conditions, the convoy depended solely on sound to stay together and act in concert. The natural sounds of the hurricane prevented important human sounds from being heard. Drums, trumpets, shouting, and the sounds of cannon and gun shot conveyed simple messages about direction and intent within a convoy. Under normal conditions, Admiral Somers "spoke" to the rest of the convoy by these means and they replied the same way. In the darkness and rain of the storm, however, the Virginia Company ships could not see each other, much less flags. The hurricane "beat all light from Heaven, which like an hell of darkness, turned black upon us." It smothered all fires. Even the cooks' stoves sheltered beneath the decks were drowned. Lanterns and cannon would not work. If the admiral and his ships were to communicate, they would have to do it out loud rather than by sight under such conditions. But Gabriel Archer, aboard one of the smaller ships, wrote that the storm was so loud that no one could "hear another speake." Drums, trumpets, and shouts could not be heard above the terrible din. One ship could not know another's place in such a "roaring sea." According to Archer, they were "thus divided" from each other by the sounds of the storm.[11]

The tempest's visceral roaring affected more empires than those of the mind. It had both immediate and long-term consequences for the survival not only of the convoy, but of Jamestown and Virginia. In the longer term, the sound of the hurricane decapitated the colony: it "separated the head from the body, all the vital powers of regiment being exiled with Sir Thomas Gates" aboard the *Sea Venture*. That ship ran aground on the shoals of "that dangerous and dreaded island, or rather islands, of Bermuda." Because of their terrible "tempests, thunders, and other fearful objects," sailors had come to call them "the Devil's Islands," in keeping with a commonplace belief that thunder was the work of demons. The rest of the convoy, the "body," foundered for nearly a week before finding each other. Only then did they limp onward to Chesapeake Bay, leaderless and paperless, for they neither saw nor heard any sign of the *Sea Venture*. The stragglers let loose "a tempest of dissension" upon their arrival. Led by Archer and George Percy, they started a mutiny and nearly took the life of John Smith, who had been the governor and de facto leader of the colony.[12] But Jamestown would not run wild; it—along with Gates and most of the *Sea Venture*'s passengers and crew—would narrowly survive. Other soundways helped them do so.

Even before 1609, sailors and settlers alike had noted the violent sounds of the Atlantic world. Christopher Columbus met thunder and lightning on his voyages that made it seem "as if it were the end of the world." Cabeza de Vaca was caught in a hurricane in Cuba in 1528 during which he "heard a great roaring and the sound of many voices, of little bells, also flutes, tambourines, and other instruments, most of which lasted till morning, when the storm ceased." Walter Raleigh's Virginia-bound fleet was held at bay by "a great storm of thunder and wind" at Plymouth (in old England) in 1583. Thunder, rain, and hail battered the frail beachhead colony at Roanoke so much that it was nearly abandoned in 1586. In 1607, even as the *Susan Constant,* the *Discovery,* and the *Godspeed* approached Chesapeake Bay carrying the first crew of Jamestown settlers, they were caught in a tropical storm that struck them all night with "thunders in a terrible manner." The thunder would not have been a surprise, though, as by that time the western Atlantic world's remarkable tempests were common lore among mariners and explorers.[13] Mariners may have gotten some of their stories about the Bermudas' thunders from indigenous Americans, who also avoided those islands. Sailors said the Bermudas were inhabited by devils because of local beliefs that the area's weather was supernaturally fearsome. What was a devil to Europeans may have been a deity or an invisible force to indigenous peoples. The word "hurricane" is derived from the name of a new world deity variously embodying storms and thunder.

Once inland, thunders ceased to be recorded as much more than an "inconvenience of the country" by the early eighteenth century.[14] During the first century of Chesapeake settlement, moderate Anglicans—who were somewhat lukewarm in their devotions—cared little for the millenarian aspects of thunder. But in the areas where more radical Protestant ways of life were practiced it still thundered marvelously.

Boanerges

The belief in agentive, powerful sounds appeared throughout seventeenth-century English—and for that matter, European—cultures. Associating these soundways with some sort of "orality" is problematic because one of the best seventeenth-century indicators of the importance of nat-

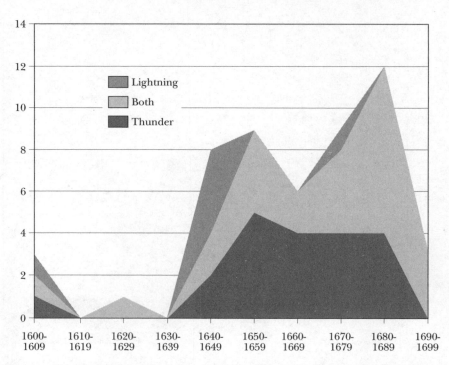

Figure 1.1 English and colonial imprints with "thunder" and/or "lightning" in their titles, 1600–1699. Richard Cullen Rath, "Worlds Chanted into Being: Soundways in Early America" (Ph.D. diss., Brandeis University, 2001), Appendix 1.

ural sounds comes from the titles of books printed and published in the seventeenth century. While the seventeenth-century imprints are all English, many of them appeared in American libraries. During the seventeenth century, there were three times as many titles containing only thunder as there were titles with only lightning. In the years between 1650 and 1690, seventeen titles appeared that contained thunder but not lightning. Only one appeared that had lightning with no thunder. When titles containing both thunder and lightning are counted, there were still three titles about thunder for every two about lightning throughout the seventeenth century. Only in the eighteenth century did publications begin to reflect the more visual ways of thinking to which Nathaniel Shurtleff reflexively translated early New England accounts of thunder and lightning.[15]

What were these books about? Some warned; others lamented or exhorted. Some were metaphorical, with thunder standing for the immanence of some military action or God's vengeance. Others explained

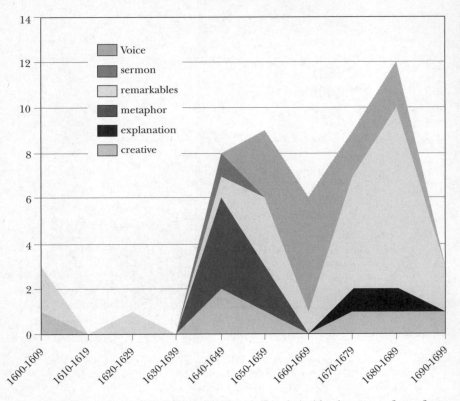

Fig. 1.2 English and colonial imprints with "thunder" in their titles, by genre, 1600–1699. Rath, "Worlds Chanted into Being," Appendix 1.

meteorological phenomena or the art of weather forecasting. Many were descriptions of "remarkable" storms. Often thunder was described as a prodigy or a wonder, and great thunderstorms that took a number of lives or struck a well-known church could generate a spate of pamphlets and books. Some titles became steady sellers, reappearing in new editions half a century or more after their original publication.

During the seventeenth century, English and American religious dissenters and separatists crafted new ways of expressing these beliefs, particularly during and after the English Civil War. They began to emphatically regard thunder as the "loud-speaking voice of God." They also intensified the idea of sound as a powerful physical force. Perhaps the best example of the Protestant emphasis on thunder as the effective voice of God came from the Quaker Robert Dingley, who wrote a book called *Vox Coeli*. The Latin word *coeli* punningly referred both to the heavens and an engraver's chisel. Thus the sound of thunder was not only the voice of the heavens; it could also work on people and things like God's chisel, with the sound physically chipping away and shaping the world. God might use thunder as his chisel, but more radical Protestants—particularly Quakers—also began writing of themselves as God's thunder chiseling away at the sinfulness of England's people and rulers. When God "sent forth his [human] Instruments, whom he had prepared and fitted for his Work," wrote Robert Barclay, their words became "like Thunder-bolts, knocking down all that stood in their Way." An anonymous pamphlet directed a "Thunder-clap to the army and their friends warning them of their imminent danger" in 1648. The fifth monarchist John Rogers warned of "Doomes-day drawing nigh, with thunder and lightening to lawyers" for the new laws of Cromwell's Protectorate.[16]

The most striking development in the mid-seventeenth-century English natural soundscape was the willingness of men, and many women, too, to claim their own voices as God's thunder. In the 1660s, Edward Burroughs, writing of the New England Quakers, had presented his own speaking voice as

> a trumpet of the Lord sounded out of Sion which sounds forth the controversie of the Lord of Hosts and gives a certaine sound in the eares of all nations and is a true noyse of a fearfull earthquake at hand which shall shake the whole fabrick of the earth . . . or, Fearfull voyces of terrible thunder, uttered forth from the throne to the astonishment of the heathen in all quarters of the eearth who are not sealed in the forehaed.

That was just in the title. A few years later, the cantankerous London Quaker merchant Humphrey Bache confronted a jeering throng that spit on him and cursed him for opening his shop on a Sunday. He railed against them in what he thought to be a divinely ordained "Voice of Thunder," calling his attackers so many wild animals in the hands of the devil. R.S. was a London Quaker who fasted and prayed and was then filled with visions and the sound of God's voice. Some time around 1660, he wrote a broadside entitled *The Dreadful and Terrible Voice of God Uttered from the Throne of his Justice as the Voice of a Mighty Thunder, and as the Voice of many Waters Rumbling*, in which he claimed that "the Word of the Lord Sounded through me." He heard God's voice as a thunder that made all quake except those who were saved. "The God of Israel" came to him "as in Mighty thunders." God "uttered his Voice and the mountains trembled, and made the hills to shake, and all the Beasts of the Field to tremble: So that none could endure his Voice, but they who were gathered into his Light; to them his Voyce was Sweet."[17]

God's Thunder could overcome volition. London Friend Dorothy White's 1661 "alarm" to the English was "Sounded" from God's throne "With the Voyce of Terrible Thunder." That voice pushed her into action, not by persuasion or reason, but by the force of its sound alone. "The Lord God hath uttered forth his Voyce," she maintained, "as a Mighty Thunder, by which I was made thus to write." She related how the same thunder that made her write had "gone forth against" England's people and rulers in recent years, equating the thundering sound of God's voice with a sword made of "his Power."[18]

"Boanerges" was translated in the King James Bible as "sons of thunder." The two apostles James and John were these Boanerges or sons of thunder because of the God-given power of their preaching. Priests and Anglican bishops might claim to be an "eccho of the sons of thunder," but Quakers and the more radical Puritans pushed the meaning to a new level. The Puritan parliament of the 1640s claimed to be "Boanerges, or the Parliament of thunder," whose proclamations struck rebels like thunderclaps. Henry Adis described himself as "one of the sons of Zion, become Boanerges to thunder out the judgements of God against oppression and oppressors." Samuel Chidley thought of his words as "Thunder from the throne of God." In the 1660s, Edward Burrough, a defender of the hanged New England Quakers, was memorialized as a Quaker "son of thunder" after he succumbed to the rigors of a London jail at the age of twenty-eight.[19] This willingness to claim one's own voice as God's thunder was new, a product of radical, literate religion and the stormy times that lay over England during the mid-seventeenth century.

The biblical "seven thunders" that were thought to be the first step in the earth's final destruction and judgment were another regular radical theme in seventeenth-century titles. In 1665, sectarian William Bayly wrote a pamphlet about the "*Seven Thunders Uttering Their Voices.*" Several years later, the Quaker Jane Lead wrote *The Revelation of Revelations Particularly as an Essay Towards the Unsealing, Opening and Discovering the Seven Seals, the Seven Thunders, and the New-Jerusalem State*. With pre-millennial fervor, an anonymous pamphlet predicted that "a most loud and publick ministration of the Gospel, beyong [*sic*] any time, since the apostles, described as seven voices, or the seven thunders unsealed." The sound of thunder, reflecting the voice of an angry God, not only could but inevitably would destroy the world.[20]

If thunder was the voice of God, and sounds were tangible forces, then it is not surprising that thunder rather than lightning was often credited with killing people and damaging property. An anonymous pamphlet told of a "clap of thunder which lately set fire to the dwelling-house of one widow Rosingrean" in 1677. Another anonymous pamphlet reported a "sad accident near Norwich" in 1680. Eight people were "struck dead in a church porch by thunder." The same pamphlet told of a man and his son who along with their four horses were "slain by the thunder and lightning" as they worked their fields. Sometimes the author was not sure what struck, as in the 1680 "account of a man living near Shorditch church, who was struck with the thunder or lightning."[21]

A rich gamut of beliefs about thunder coexisted in the publications of seventeenth-century English writers. Many of these sonic beliefs made the passage across the Atlantic. In Puritan New England, which saw itself as carrying on the Puritan experiment after Cromwell's protectorate had failed in England, God's voice was heard loudly in cracks of thunder during the second half of the seventeenth century.

New England's "Terrible Cracks of Thunder"

Protestant natural soundways traveled to Puritan New England and thrived in the midst of the most literate society in the world at the time. If literacy—according to proponents of the literacy theory—brought about a more visual orientation to the world than it did to "pre-literate" or "oral" "people of the ear," then it would stand to reason that Puritan New Englanders should have emphasized lightning, leaving little power to the sound of thunder. Yet in seventeenth-century descriptions of New England storms, "thunder" or "thunder and lightning" were often held accountable for the worst of the physical damage.

Increase Mather began a discourse on lightning with a remark that "lightning and thunders were frequent in this land, yet none were hurt thereby" in the early days of the colony, but for the later years of the seventeenth century he wrote "an account of remarkables respecting thunder and lightning" that did hurt and kill. He transposed the order between the two uses of the terms—benevolent "lightning and thunder," but deadly "thunder and lightning."[22] When the thunder was the primary term, the damage was fatal. More often than not, it was a clap of thunder that killed rather than a stroke of lightning.

From about 1650 to 1690, New Englanders adapted and extended English Protestant soundways as part of the Puritan colonial experiment. Published English and continental examples of deaths and accidents by thunderclaps were well known, and often discussed, as when "the Main-Mast" of a French ship "wes split in pieces with a clap of Thunder," or when another ship's compass was knocked out by "one dreadful clap of thunder," or when a man walking in an English field was killed "with a Clap of Thunder."[23] Mather actively sought such stories on both sides of the Atlantic and wrote two chapters of his *Illustrious Providences* on thunder and lightning, peppering the remaining pages liberally with more remarkables concerning thunder. His transatlantic connections marked him as an elite in some ways, but they also let him connect his book with popular English lore that resonated with New England commoners, making *Essay for the Recording of Illustrious Providences,* or *Remarkable Providences* as it came to be called, a best seller.

People discussed such stories for entertainment, sometimes generating new wonders in the process, as when Constance Southworth of Duxbury came home from evening military exercises on Sunday, September 11, 1653. A storm was brewing, though until evening it consisted only of drizzle and "some seldome and scarce perceivable Thunders." The distant sounds set the men to "Discoursing of some extraordinary Thunder-claps with Lightning, and the awful effects and consequences thereof." He took leave of the men and went into his own house, where his wife, two children, a servant, and the family dog were. Immediately upon entering, "there broke perpendicularly over the said House and Room a most awful and amazing clap of Thunder, attended with a violent flash, or rather flame of Lightning," which sent pieces of the chimney flying and filled the room with "Smoke and Flame." It melted pewter, set the roof of the lean-to on fire, struck some of the people in the room, numbing them or knocking them windless for a few moments, and killed the dog literally in its tracks: it died "giving a small yelp." Remarkably—and that was the point—no one but the dog was killed.[24]

Before the last decades of the century lightning would seldom be considered the sole cause of death. Mather recounted a 1654 storm of "thunder and lightning" that set fire to the house of a Salisbury man by the name of Partridge. While Partridge was outside fighting the fire, he was hit "by a second crack of thunder with lightning" by which he "was struck dead," and—underscoring the connection between animacy and sound—"never spake more." This use of redundancy is today often considered to be a mark of orality, a trait not usually associated with Mather's voluminous written output. In the first sentence, in which a fire was started, Mather used the conjunction "and," indicating coordination, a grammatical form marking relative balance between the words "thunder" and "lightning." In the second lethal stroke, however, Mather subordinated lightning to the thunder by means of word order, but more important, through the subordinating conjunction "with." Thomas Morton wrote of a 1658 storm that "by Thunder and Lightning one *John Phillips* of *Marshfield* was suddenly slain." Increase Mather reported the cause as "a terrible crack of thunder." At the inquest that followed, a jury "swore" that it was "by an Immediate hand of God manifested in Thunder and lightning [that] the said John Phillips came by his death."[25] Again, thunder was mentioned first (and, perhaps significantly, capitalized), and the importance of sound was accented by the jury's swearing aloud before the deity.

Obviously, people were aware of the visual aspect of thunder and lightning, and a few examples record something other than thunder as providing all or much of the potency of a strike. "In a storm of thunder and rain" at Northampton in 1664, a "ball of lightning" struck and killed Matthew Cole as he stood in a room of about a dozen people. In 1665, a Captain Davenport "was killed with lightning" as he lay sleeping in Boston. In his diary that day, however, Mather recorded that he was "affected by news of Capt Davenports being killed with thunderbolt yesterday." Nearly twenty years passed between Mather's initial response and Davenport's death "with lightning" in *Remarkable Providences,* perhaps reflecting a shift toward the visual at this later date. Drawing on radical sectarian language, the Quaker George Keith claimed that Davenport's death was from the " sounding of God's Voice from heaven," exacting retribution for Davenport's persecution of Quakers.[26] Even in these counterexamples, the role of thunder is often greater than it would be today. Though aware of lightning, they had a greater sensitivity to the auditory than do we.

Usually, it was the "terrible cracks of thunder" that inflicted the most damage. Describing the onset of the 1666 Marshfield storm, Nathaniel Thomas wrote that "the Thunder came quickly up over the house." Then, "an astonishing thunderclap fell upon the house," claiming three lives, including that of the elder John Phillips, father of the 1658 victim. William Hubbard wrote to Mather that the three were "suddenly killed in a storm of thunder." John Josselyn also described them as being "kill'd in a moment by a blow of Thunder." Thunder did property damage, too. William Jones wrote of his New Haven colony home: "That which befell my house was by a dreadfull thunder clap and lightning."[27]

Often, those who were struck by lightning but spared the sound of the thunder lived. "Divers" others who were merely "burnt with lightning" during the second Marshfield storm survived. The wife of the elder John Phillips was "preserved, though in some measure scorched by the lightning." Also in 1666, Samuel Ruggles of Rocksborough was "struck with Lightning. He did not hear the Thunder-clap, but was by the force of the lightning, e're he was aware, carried over his Cattle about ten foot distance from them." Perhaps the author was implying that had Ruggles heard the thunderclap, he would not have lived. As it was, Ruggles recovered fully, being only temporarily paralyzed and a little scorched on the legs. At the house of Thomas Bishop, again at Rocksborough, "lightning" melted some dishes, but no one was hurt. In 1673, off the coast near Marblehead, "lightning" mentioned alone injured several but killed none immediately. With the thunder, death occurred, as at Wentham in 1673, when "a smart clap of thunder broke upon" a Mr. Newman's house. "With the Thunder-clap," wrote Mather in 1683, "came in a great ball of fire as big as the bullet of a great gun." It traveled about the room, killing Richard Goldsmith, a shoemaker who had allegedly been prone to leaving jobs half-done.[28]

Thunder could effect changes in the marrow of the souls of New England's women and men. The effectiveness of one of Mather's most famous sermons, *The Day of Trouble Is Near,* turned on his evocation of God's thunder at the crucial moment, manifested in roaring earthquakes and Sion's trumpets. In this archetype for the jeremiad tradition, Mather used sound to frighten and push his audience toward repentance and piety.[29] But alas, even as it was reaching its peak, the sound of thunder was beginning to lose its sway over New Englanders.

During the 1680s, Increase Mather attempted to reconcile new academic theories of thunder and lightning with his belief in the active intercession of providence in this world. He began with a statement that thunder and lightning were "very mysterious and beyond human capacity to comprehend." He substantiated this claim with biblical quotations, all concerning thunder alone: "the thunder of his power, who can understand? . . . Can any understand the spreadings of His clouds or the noise of His tabernacles? . . . from Job, 37:5: He thundreth marveils [*sic*]." Although thunder was considered to be the very "voice of God," Satan and particularly malevolent popes were also believed capable of causing it. In this spiritual voice, he was working in the same genre of "remarkables" and "wonder stories" that comprised the perennial best-sellers of English thunder and lightning books between 1640 and 1690. In fact, he had pursued a concerted and transatlantic effort at collecting such "remarkables" that the Puritan divine Matthew Poole had initiated in England a few years earlier.[30]

Mather did provide a scientific explanation in *Remarkable Providences.* He considered lightning to be "a downward-burning fire of nitre" and other materials, including sulphur and brimstone. He thought it a commonly held vulgar error that stones came down with the thunder, noting that thunder was often not heard by those struck. He knew that the clap heard was not the fatal one. How it killed remained a mystery to him, however. Some contended it was through vapors, but he held it was more than that— it "Lick[ed] up the vital spirits that run in the body, the *vinculum,* the tye of union between the soul and the body."[31] Thus thunder could still take the very life and soul out of a person in Mather's scientific explanation. He had an educated hunch that it was not the sound that did it, but no alternate explanation, and he was not in the least hesitant to turn back to folk and spiritual beliefs in the power of sound to make his point, even when those beliefs directly contradicted his academic explanations.

As early as 1686, Charles Morton would be teaching his Harvard classes that earthquakes, thunder, and lightning were purely natural occurrences with no portentousness at all. According to Morton, rain clogged the earth's pores so that the underground nitre and sulfur could not be vented except through the plants they helped to raise. When the sun came out, evaporating the water, the water vapor carried the nitre and sulfur with it into the heavens. "These steams," thought Morton, "get away to become the matter of thunder," combining in the clouds to produce a sort of gunpowder. The actual damage, according to Morton, was done not by the thunderclap, but by a "bolt" not unlike a bullet that was a "meteor of the middle region" formed of sand and dirt brought up into the clouds with the vapors. Earthquakes in turn were nothing but thunder underground. Even so, natural sounds could do physical damage, as when Morton explained that the sound of thunder could spoil beer by "the Smart percussion of the Air" that "causes all the vessels and liquors to Vibrate which Does so allter the site of those minute, and volatile parts, (that preserve the liquor[s] by their ordinary circulation) as that they can't do their Office; and the Same concussion also gives motion to the Corrupting parts which before ware Quiet in the Lees." The same effect could be achieved with any loud noise, including bells or guns, he believed.[32]

Neither thunder, earthquakes, nor spoilage was in any way supernatural in Morton's account, although he still gave a mechanical explanation of some of sound's powers. It would be a long time before that teaching trickled down from Harvard graduates to the rest of the population, though. In 1683, when Mather's *Illustrious Providences* was published, Morton had written "Compendium Physicae," but he would not migrate to Boston and teach at Harvard for another three years. Mather, as the key New England member of what Francis Bremer has called "the trans-Atlantic congregational network," knew that something was in the winds, and did not want his explanations of thunder to appear dated, but he was also trying to reach a less-educated audience than was Morton.[33]

The tension between older and newer natural soundways is evident in Mather's attempt to elide inchoate academic beliefs with old spiritual ones. Puritan beliefs about sound would linger on, ever more in the background, until well into the eighteenth century, while at the same time visual modalities came to the fore. In hindsight, and at first glance, this coexistence of seemingly incommensurate sensory regimes seems superstitious and irrational. Once we give the Puritan ways of understanding the sounds about them a hearing, however, we begin to make out complex, polyvocal, and sometimes dissonant ways of making sense of the natural world as they existed in the flux and flow of history, even appearing to clash and contradict within the writings of a single author. With the rise of new media in the past century, we may be in yet another era of contradictory consciousnesses, so they might be worth a closer look—and listen.

Thunders Underground

Puritans described earthquakes in terms of the sounds they made, too. A 1638 tremor "was heard before it came with a rumbling noise, or low murmur, like unto remote thunder. . . . As the noise approached near, the earth began to quake. . . . About half an hour after, or less, came another noise and shaking, but not so loud nor so strong as the former." The author then remarked with wonder on the power of God. Roger Williams called it God's "late dreadfull voice and hand: That audible and sensible voice, the Earthquake." Mather noted that the same earthquake was said to have been preceded "by a strange kind of noise before the earth began to tremble."[34]

Earthquakes were interpreted by all as providential wonders, so it is not surprising that they should be preceded by otherworldly sounds. When one struck Essex County in 1685, Samuel Sewall reported that "that which most was sensible of was a startling doleful Sound," which someone in Portsmouth described as having "a fine musical sound, like the Sound of a Trumpet at a distance. He could not distinguish any tune that he knew; but perceived a considerable variety of notes."[35]

The attribution of musical notes at first seems like too much imagination, but geologists have established that weaker earthquakes produce a higher-pitched noise than stronger ones. The 1685 quake was probably a very weak one, perhaps not even felt but only heard. Rather than a deep rumbling, the milder quake could have been a higher-pitched sound like a trumpet. Joshua Moodey reported another quake in the summer of 1688. He had heard about it from Nathaniel Byfield, a proprietor and early settler of Bristol, Plymouth. Byfield was "an earwitness of the same" earthquake, making him a credible source. At York and Casco, another mild earthquake may have been registered when "at sundry times many gunns have been heard," though none of the neighboring towns admitting to firing any. Moodey concluded "they were the noise of Guns in the Air, which must needs bee reckoned awfully prodigious."[36]

The site of the present-day town of Moodus, Connecticut, was known as "the place of noises" to seventeenth-century inhabitants of the region. The Moodus noises were real, and occasionally recur to this day. Current geological theories hold the noises to be the result of seismic activity. Clusters or "swarms" of tiny earthquakes vibrate the Earth's surface; then the vibrations are carried into the air and heard as thunder and banging sounds. Geologists believe that these were tiny quakes with their epicenters deep underground below Moodus.[37]

In line with European American soundways concerning thunder, Puritans interpreted the Moodus noises ambiguously. On the one hand, they could be God's "dreadfull voice." On the other hand, they echoed sailors' explanations of why North America had so much thunder when they argued that the Moodus noises resulted from it being "a place where Indians drove a prodigious trade worshipping the devil" in the years before the English invaded. The English did not make any serious settlement there until the 1730s, perhaps in part because of the noises.[38]

Stephen Hosmore left the most thorough colonial record of the Moodus's rumblings. His description is a combination of an oral history that recounted local beliefs from times prior to English settlement and his own "earwitnessing" of the sounds in 1729. Of the latter, he wrote:

> I myself have heard eight or ten sounds successively and imitating small arms, in the space of five minutes. I have (I suppose) heard several hundreds of them within this [past] twenty years, some more, some less terrible. Sometimes we have heard them almost every day, and great numbers of them in the space of a year. Oftentimes I have observed them to be coming down from the north imitating slow thunder, until the sound came near or right under, and then there seemed to be a breaking, like the noise of a cannon shot, or severe thunder, which shakes houses and all that is in them.[39]

John Bishop, a minister in Stamford, Connecticut, wrote of "the noise of a great gun in the air at some of the Norwootuck [Norwalk?] plantations" and "the noise of small guns in the air" heard northeast of Stamford (Moodus is just to the northeast, but was not named). He then reported that there had been an earthquake that he had been "sensible of" early in 1678. A few months later "a like noise was heard here by myselfe and many others, who took it to be an earthquake, rather than thunder, considering the circumstances though the terræ-motion [was] not so perceptible." He was writing to Increase Mather partly for advice, for "the awful workes of God are so variously and uncertainly spoken of; as many times I find that we know not what to believe, nor how to bee affected as we should with what we heare."[40]

The Songbirds' Elders: Native American Natural Soundways

"Moodus" is a clipped form of the Wangunk term *Machemoodus* (other spellings include *morehemoodus* and *mackimoodus*). To the Wangunks, *Machemoodus* was a place of powerful maleficent spirits who were embodied in the subterranean sounds they made. The Algonquian language of the Wangunks has been lost, but seventeenth-century lexicons from nearby communities with related languages have a word *machees* or something similar that refers to an unfriendly or malevolent spiritual power, probably a kind of underground thunder, not so far from Puritan beliefs. The place name bears out a widespread Algonquian and Iroquoian belief that sounds from below were often malevolent forces, while those from above were more protective.[41] Puritans noted that Indians worshiped this sound to appease it rather than out of love for it.

In 1638, Roger Williams wrote that older Indians remembered five occurrences of the Moodus noises, which he also identified as earthquakes, since about 1570.[42] The source of these sounds was a bone of contention between Puritans and Indians. Both groups believed that the noise was the work of animate beings. The English claimed that the Indians invoked some sort of demonic force. To the Indians' ways of hearing, however, the English had things backwards. Machemoodus inspired their worship; their worship did not conjure the sounds. The basis for this difference lay in their respective religious beliefs. For Puritans, their God did no evil and was the only appropriate power to worship. Indian beliefs, by contrast, acknowledged evil as a power that required worship and appeasement rather than demonization.

Indians undermined European religious beliefs with alacrity. Hosmore was the first minister of East Haddam (which lay just south of the center of the Moodus noises). In a letter, he remembered being "informed that (many years past) an old Indian was asked what was the reason of the noises in this place, to which he replied that the Indian's God was very angry because [the] Englishman's God was come here."[43] It is important to note that the old Indian's explanation did not treat the sounds as the voice of a deity.

Wangunks and other Native Americans attributed acts of identity—that is, they attributed the practice of attributing identity—to all natural sounds in such a way that those sounds could be coherently explained in their worldview.[44] The explanation of a powerful natural sound was that it was performing some act of identity, not necessarily in relation to those who heard it. Because such sounds were potent forces, understanding the nature and meaning of their acts could be a determinant of how well one did in the world: it sometimes explained how the thunders expected people to act. Other times, the thunders were concerned with matters that had nothing intentionally to do with humans, but nonetheless affected them, such as ball playing, hunting, or carrying in a change of season. Native Americans thus thought it necessary to perform their own acts of identity in response to those they attributed to the natural sounds, creating a reciprocal belief system out of the sounds and their responses to it. This distinction between acts of identity and speech acts becomes clearer once Native American soundways around thunder and other natural sounds are examined.

Throughout the Americas, from the Arctic to the southern reaches of South America, First Nations cultures conceived of storm phenomena, particularly thunder, as the markers of powerful beings acting on earth. They thought that these beings inhabited tempestuous domains such as mountains—or in the Caribbean, the Bermudas.[45] Such beliefs, and the practices they engendered, provide a way to reconstruct parts of Native Americans' natural soundscapes during the seventeenth century.

At the most general level, Native Americans' natural soundways corresponded to those of Europeans in several ways. Both had room for the belief that all sounds had animate sources. Both treated natural sounds as bridges between visible and equally real invisible worlds. Thunder and lightning again make a good case study, because Native Americans, like Europeans, were very much concerned with storm phenomena, and there is a relatively rich documentary and ethnographic record of their beliefs and practices. Both Native Americans and Europeans thought of thunder as a sound made by some great spiritual being. Both Europeans and Indians gave the sounds of nature a greater weight in proportion to its sights than would nineteenth- and twentieth-century academic observers.

Distinctions between First Nations and English soundways abounded. Seventeenth-century Native American practices concerning natural sounds are no more easily reduced to some generic "oral" or "primitive" culture than were European beliefs. In Native American soundscapes, the making of a sound was first and foremost an act of identity, no matter what the source. If the sound came from a clap of thunder, then an identity was attributed to it just the same as if the source were a gesture, a speech, or a song performed by a neighbor. Scholars call this "animism," but, as we have seen, Europeans were equally animistic in this regard, for they too attributed personalities to thunder and other natural sounds. Polytheism distinguished First Nations animism from the European version, so that similar soundways present in both cultures generated remarkably different expressions. The distinction that we now make between "animate" sounds made by living things and the sounds of inanimate phenomena like earthquakes and thunderstorms and waterfalls made no sense in a Native American world where sound served as proof of the presence of a living being with a particular identity. Sound was a distinguishing feature of living beings, so this was not necessarily pantheism (the belief that all the universe is alive): sound itself was alive. For example, when the Jesuit missionary Paul Le Jeune asked his Huron hosts whether the seasons were men or animal beings, the Hurons replied that "they did not know exactly what form they [the seasons] had, but they were quite sure they were living, for they heard them, they said, talking or rustling, especially at their coming, but could not tell what they were saying."[46] Contrast this with seventeenth-century European soundways, where the cause of the sound, rather than the sound itself, was what was alive. This distinction resulted in the boundaries between "nature" and "human" and between "animate" and "inanimate" being cast differently in seventeenth-century Native American and European cultures.

Four distinctive aspects mark seventeenth-century Algonquian and Iroquoian natural soundways. First are the general properties that Native Americans attributed to thunder. The second shows how Native Americans constructed identities from the sounds that they heard by looking at the forms in which they put thunder. The third part shows ways that Native American cultural groups then situated their own identities relative to those they constructed for thunder and other natural sounds. The fourth part discusses a closely related topic, the significance of the sound of waterfalls.

What were the properties of thunders? First, they were plural. There was no single thunder god. Second, as it was for Puritans, the sound was the source of the power that we now generally attribute to lightning. "Thunders" caused lightning in Algonquian and Iroquoian accounts. "Thundering" is the English translation for any storm in Ojibwe, an Algonquian language. There, according to the ethnographer of religion Theresa Smith, thunder "both rules and defines the storm." The word for lightning is seldom used to describe the phenomenon in general or as a whole; lightning is a particular aspect of the thunder. The sound of the thunder is the generic: thunder is a cause, the lightning an effect. In Cherokee mythology, the thunders used lightning as a tool. The thunders gave it to people for fire, or used it for hunting. The sound of thunder rather than the sight of lightning was the indicator of identity.[47] For the purpose of explaining causes and effects, the thunders' sound was thus a better source of data than the flash of lightning.

The Iroquois used thunder metaphorically to stand for a great military power in their treaties with the English. In 1694, the Onandaga sachem Sadakanahtie implored the English as part of a treaty between them to "Do you but your Parts, and Thunder itself cannot break our Chain." Another treaty between the Mahicans and the English in 1720 called their agreement a "Chain of Silver that the thunder it self cannot break it." Not only was thunder a tangible force, it was one of great magnitude.[48]

A clap of thunder was the act of a specific being, a personal act that helped establish the identity of its source, a "thunderer," who generally had the ability to fly. There was not a single mythical thunderer, like Thor the god of thunder. "Thunderers" flew in small groups and each clap came from a particular one. Thus the plurals "thunders" and "thunderers" fit Algonquian and Iroquoian soundways better than "thunder." Smith points out that these beings both made and were the thunders: they embodied it, and the thunder was their presence. The Ojibwe cover term for thunder is *animikeek*. But it marked an aggregate or ensemble, not a generic phenomenon. Just as a ship's crew comprises the captain, the mate, sailors, a surgeon, a trumpeter, and so forth, so *animikeek* comprised many types of thunderers: *ninaminabides* was the overseer, *nigankwan* was the first thunder, *beskinekhwam* was a thunder that hits something, *anjibnes* was a thunder that renews, *besreudang* was an echoing thunder, *bodrendang* was an approaching thunder, and *bebomawidand* was a searching thunder that advances and retreats, perhaps like a scout.[49]

Thunders had more to their personalities than the general qualities described above. They responded situationally. As a characteristic of identity, the sound of the thunders was not thought of as a byproduct of generic forces. Nor was it the effect of a personality as in European American soundways. The sound itself was thought to be intelligent. Le Jeune noted in 1637 that the Hurons would leave their javelins pointed upward outside their cabins because "as the thunder had intelligence," it would take care to avoid the sharp weapons, thus protecting the cabins.[50]

Europeans were aware of how different this notion of thunder was from their own. The Jesuit missionary Jean de Brebeuf, writing in 1628, was exasperated with a drought- stricken Huron community that claimed he was scaring away the thunders, here conceived of as birds, with his religion. The Jesuit chided them (and, perhaps more to the point, his French readers) that "only an ignorant person would say that the thunder is afraid; it is not an animal, it is a dry and burning exhalation which, being shut in, seeks to get out this way and that."[51]

Though natural sounds were acts of identity, First Nations people did not necessarily conceive of these sounds as intentional (as did seventeenth-century Europeans). The sounds of thunderbirds' wings flapping, or of the thunders playing ball in the skies, were accidental. Human deaths and damage from what we would now call lightning strikes were attributed to the wildness of youthful thunders as often as to any intentional act of retribution.[52] Thus Native Americans understood thunder as a sign of the presence of an otherworldly being rather than an ominous voice in dire need of interpretation. The thunder was not necessarily about them, and even when it was, Native Americans construed thunders as protective forces rather than as the effect of a retributive, monotheistic force.

Thunderers—those who actually made the sound of thunder—seem to have been mostly males. Historical sources, when they mention gender at all, tend to use male pronouns, as when Brebeuf reported in 1636 that the Hurons thought the thunder was a man in the form of a bird. This may have been due to the grammatical genders of the French and English languages rather than from gender as constructed by the Hurons, however. Modern ethnographies do ascribe distinct gender roles to the thunders, but they may be unreliable for the seventeenth century. In modern accounts, thunderers are most often, though not exclusively, male. Ojibwe people today consider thunder to be manly, thus construing women who thunder socially as male. In most ethnographies, women are the daughters or mothers of the thunderers. While this makes them female thunders, they are not the ones who make storm sounds. Actual thunderers sounded off in the pursuit of predominantly male activities like hunting and ball playing. Similarly, western Apache stories distinguish male rain from female rain: the former is with thunder, the latter without it. But Eskimo stories tell of thunder girls whose rattle-playing was the source of thunder; thus comparative approaches do not increase our certainty about how natural sounds may have been gendered in the seventeenth century. All that can be said is that thunder was associated with male social functions in many places.[53]

The thunders were great possessors of what was called *manitou* in Algonquian languages and *orenda* in Iroqouoian languages. Roger Williams demonstrated the meaning of the concept in the 1630s, reporting that "There is a generall Custome amongst them [Narragansetts], at the apprehension of any excellency in Men, Women, Birds, Fish, &c. to cry out *Manitoo,* that is, it is a God." Iroquoian beliefs were similar. Neil Salisbury interprets this recognition of excellency as a "manifestation of spiritual power, a manifestation that could occur in almost any form." Manitou did not float around freely: it always took some form or another. While the power was impersonal, it could not be expressed without somehow being embodied. The manitou of these forms was often recognized in the sounds they produced. As with Europeans, spiritual power was closely associated with the sounds of nature. In the seventeenth-century Algonquian and Iroquoian world, living beings were the source of all sounds, so thunder was by necessity a live entity.[54]

In summary, Native Americans attributed seven properties to the thunders. Thunders were tangibly powerful. They were responsive to earthly signs. They were reasonable and intelligent. They could feel emotions. Their concerns were not always human-centered. They may have been predominantly male. They possessed powerful spirits.

What forms did thunder take? Native Americans often described thunder as the sound coming from the wings of great birds. The Huron shaman Onditachiaé maintained that thunders took the form of "a man like a Turkey-cock." The rumble of thunder came from flapping wings of such men. In 1636, he explained to Brebeuf that "if the uproar is a little louder, it is his [the thunder-being's] little ones who accompany him and help him make a noise as best they can." Le Jeune wintered with an old Montagnais Indian in 1633, whom he had quizzed on the subject of thunder. The old man claimed that the Montagnais "did not know what animal it was," but that "the Hurons believed it to be a very large bird." Brebeuf corroborated this a few years later when he wrote of "the thunder, which they [the Hurons] pretend is a bird." Le Jeune's informer told him that the Hurons "were led to this belief by a hollow sound" made by a "kind of swallow" that flew around on summer evenings "repeatedly making a dull noise." Le Jeune maintained that "the Hurons say that they [the actual birds] make this noise from behind, as does also the bird which they think is the thunder." "From behind" may have indicated some kind of flatulence, but more likely, since that explanation turns up nowhere else, it meant that the sound came from the birds' wings, as opposed to its songs or cries. Although Le Jeune discounted the belief as a devilish superstition, he also mentioned examining one of the smaller birds as part of his inquiry into Huron accounts of thunder, indicating he took the belief seriously enough to require his experiment.[55]

Thunderbirds fit into seasonal First Nations soundways and the migration patterns of smaller earthly birds. Hurons observed that both followed the same seasonal path. Songbirds arrived from the south in the spring, these smaller birds making smaller sounds that foretold the arrival of their "elders," the thunders, later in the season. Both the songbirds and the thunders departed with the end of summer. The thunders lived in the mountains, nesting above clouds that obscured them from sight, much like their nearest rivals in size, the eagles. The thunders, conceived of as much greater than eagles, lived in even higher and more inaccessible places.[56]

Many Native Americans conceived of the thunders as human in form. Nineteenth-century Algonquian and Iroquois ethnographies recount "thunder stories" in which the thunderers were much like Indian men and women except that their villages were in the mountains or the clouds. These thunderers had removable wings that they would occasionally don, and their flapping made the sound of the thunders when they played or hunted.[57]

Thunders were beings by dint of the sounds they made rather than the forms they took. The lines between human and deity and between human and animal forms were permeable and amorphous. A thunderer could be in the form of a bird one moment and like a human the next. "Beings" is the only general way to describe them, for it was not clear always whether their form was corporeal, spiritual, or both at any given moment.

What were the relationships between humans and thunders? Men and women were said to live with thunders, procreate with them, and sometimes even to return to their fellow humans to tell the tale. Native Americans situated themselves and the sounds of thunder within a complex social web. By attributing acts of identity to thunder-beings in the sounds they made, Native Americans were able to situate their own identities in terms of their relationships to their natural soundscapes. The relations a person had with various manitou-bearing sounds could determine success or failure in life.

Human relations with the thunders were extensive and varied in the seventeenth century. In the 1630s, the Huron shaman Onditachiaé claimed he could control the rains because of his special relation with the thunders. Brebeuf did not dismiss Onditachiaé's claims. He believed that the shaman's powers were real, but derived from devils. Shamans had too long a record of success for their powers to be imaginary, Brebeuf thought, yet it was unimaginable to him that they could have derived their powers from God. His fellow missionary Le Jeune was also inclined to believe that the devil was directly involved in Huron and Iroquois religion, not that shamans merely pulled tricks.[58]

A few years later, Father Louis André wrote of a shaman who "had an exceedingly great confidence in thunder as a powerful divinity; and, far from hiding when he heard it rumble, he did all that he could to meet it." André witnessed the shaman one day during a thunderstorm as "he ran about in the woods, entirely naked, crying aloud and invoking the thunder by his songs." The shaman's purpose was, in André's words, "to lead to the belief that he was seized with an extraordinary enthusiasm, of which the thunder-god was the author." André told the shaman "he had reason to fear lest God, who uses [thunders] as a hunter uses his gun, should discharge it at him, and make him die instantly." This apparently convinced the shaman to cease his practices.[59]

Thunderers visited people in dreams. Father André and other missionaries complained that Native American attention to dreams was a major obstacle to their attempts at conversion. Native Americans in his mission near present-day Green Bay often claimed that thunder spoke to them in their dreams. The suspension of earthly rules in dreams helps explain some of the seemingly fantastic beliefs they held. People returned the thunders' visits, not only in their dreams, but in the telling of stories about the thunders. Algonquian and Iroquois stories still tell of people who became thunderers for a time, many of whom would then return to their human villages either seasonally or after a long absence. Native American dreams and stories both inspired and validated such expeditions.[60]

In Iroquois tales, the chief thunderer, Hinon, had an intimate relationship with humans, even siring children by human mothers. Once old enough, Hinon would give these children wings and they would then fly off each season with him to make the thunders. Similar stories of human-thunderer intimacy appear in Cherokee, Huron, and Passamoquoddy tales. In some, a hunter or a boy stumbles upon the home of the thunderbirds; in others, it is a woman who encounters them. In the stories of female encounters, the woman is often supernaturally pregnant and has either left or been cast out of her village. In one nineteenth-century Iroquois variation, she throws herself over Niagara Falls, only to be caught and taken in by Hinon, who in this variation lives under the falls, causing them to roar. The woman is asked to dance, and in doing so gives birth to a number of serpents that Hinon kills, making it possible for her to return to her village. While these stories are suggestive of attitudes about sex and gender, disentangling modern embellishments from older themes is impossible without corroborating documentation.

One seventeenth-century theme is of thunder benevolently protecting Native Americans from an antagonistic serpent, or sometimes a giant horned worm or caterpillar. Onditachiaé maintained that the thunders hunted snakes and "all they call *Oki*." Brebeuf defined *oki* as anything that gave one power, or the being that held that power. Le Jeune also reported that thunderers hunted and ate great snakes. During a storm in 1637, Father Buteux asked one of the Hurons where the last great clap of thunder had come from. The Huron replied that it was "the Manitou who wishes to vomit up a great serpent he has swallowed; and at every effort of his stomach he makes this great uproar that we hear." The flashes of lightning were the great serpents falling to the ground. Theresa Smith argues that in the modern Ojibwe cosmology, people live on an island in a vertically oriented world, in a tense dialectic between the protective thunders who hunted a great malevolent serpent below the waters. While the thunder may have been a particularly important part of the Ojibwe cosmology, where she did her fieldwork, Smith draws her conclusions from Algonquian and Iroquoian evidence too, so her conclusions can be applied more broadly.[61]

Though benevolent, thunder was also to be respected and even feared. Strachey, after recounting the Bermuda storms, reported that the Powhatan people's deities marked their displeasure toward the Virginia Indians by thundering. He did not dismiss this belief as idle superstition. Instead, he reported that the Powhatans were thralls to powerful devils. He thought that the Indians' pagan religion caused the "devils" that they worshiped to bring so much thunder and lightning down on the colony as was not "either seen or heard in Europe as is here."[62]

The idea of thunder as the voice of a deity, so common in the seventeenth-century European American soundscape, does not emerge from the seventeenth-century Native American historical record. The combination of the incomprehensibility and animacy of natural sounds was not limited to thunder. Although the seasons spoke in the rustling of the wind, seventeenth-century Hurons were careful to note that it was not articulate to them. Likewise, the female spirit responsible for diseases could be heard in campfires, "roaring like a flame but her language cannot be understood."[63]

First Nations people claimed, perhaps facetiously sometimes, that the sounds of waterfalls were animate, too. The Menominees discouraged the Jesuit Jacques Marquette and the explorer Louis Joliet from their pioneering trip down the Mississippi River, saying "that there was even a demon, who was heard from a great distance who barred the way and swallowed up all who ventured to approach him." Marquette scoffed, and he proceeded down the Mississippi in 1673 with Joliet and his company. But just above the confluence of the Mississippi and Missouri Rivers, the skeptics heard the roar. At about the same time, they sighted two "monsters" painted on some high rocks which "at first made Us afraid." The sound of the "demon," it turned out, was the noise of the rapids formed by the Missouri emptying into the Mississippi. The Recollect friar Louis Hennepin reported that the falls at Niagara "thunder continually." The falls themselves were quiet, he claimed, "but when this great mass of water reaches the bottom then there is a noise and a roaring greater than thunder." Under the falls, where Hennepin heard "those thunders, those roar-ings," was where the Senecas placed Hinon, the chief of the thunders.[64]

Thus it is no surprise that Iroquois people and other Native Americans paid close attention to the sounds of waterfalls. They might have been able to hear them better than could Europeans, much like a musician can pick out a particular instrument from a mix of sounds where a nonmusician might hear everything together. Although probably an exaggeration, Hen-nepin's claim that seventeenth-century Iroquois could hear Niagara Falls from a distance of fifteen leagues supports the presence of this skill at some level. Writing in 1750, Peter Kalm reported that the Senecas paid close at-tention to variations in the sound of the falls. A loud day was a "certain mark of approaching bad weather, or rain," wrote Kalm, and "the *Indians* here hold it always for a sure sign."[65]

In one case, Native Americans constructed part of their identity from the sounds of Niagara Falls. From the mid-1630s to the 1650s, the Five Nations and their western neighbors were caught up in a cycle of violence driven by disease and the fur trade. In 1640 and 1641, smallpox epi-demics struck the Senecas and the Neutrals who lived to their west near Niagara Falls, killing up to half their populations, particularly elders. Seneca stories record this as an allegory. A young woman, representing the Neutrals, was in a state of despair over the losses from disease and was about to throw herself over the falls. Hinon—the chief of the thunder-ers—lived in the falls as the embodiment of Niagara's roaring sound. He interceded, explaining to the woman that the disease came from a great snake that lay under the falls and poisoned the water. As elsewhere among Iroquoian-speaking peoples, protective thunderers hunted malevolent snakes. Hennepin—who left the first written report of the falls in the late 1670s—corroborates this story, missing the allegorical aspects. He re-ported that the Neutrals abandoned their settlement in part because the falls were infested with poisonous snakes. Next, Hinon instructed the woman to tell the Neutrals to move away from the falls so he could kill the snake, which he did by throwing it onto the falls, where its body became what we now call the Horseshoe Falls.[66]

The Neutrals did in fact move away from the falls, although scholarly ac-counts offer a different version from the folktales. Standard histories say that the Senecas and their Five Nations brethren responded to disease with wars of mourning to revenge the dead and replace their losses. First they struck against Hurons to the north. In doing so, they also hoped to gain direct access to the rich northern fur areas beyond Huronia. Around 1650, after scattering the Hurons, the Senecas, with help from the Mohawks, attacked the Neutrals, who lived at Niagara Falls. The Senecas supposedly defeated them, scattering the survivors.

The story of Hinon and the young woman may have had more layers, though. Hinon represented the Senecas' and Mohawks' military strength. Remember that in Iroquois treaty language thunder was thought a great power that only the best alliances could withstand. This knowledge indicates that perhaps the Neutrals were brought in through negotiations or threats rather than by military conquest. Those that chose not to submit to the Senecas were then perhaps attacked and scattered. We know that some Neutrals fled North to the Hurons, while others went eastward, where they lived as a distinct western subgroup of the Senecas for many years before finally being assimilated. In the folk version, however, the Neutrals who joined with the Seneca came under their protection rather than their assault. The Senecas, represented by the roaring of the falls embodied in Hinon, instructed the Neutrals to get out of the way so that they could have unfettered access to the northern fur trade. Hinon's thunder strikes—embodied forever in the sound of the falls—may have represented the Seneca's benevolent (from a Seneca perspective) removal of the Neutrals and their protective adoption of the Neutrals who became the source of the story. The disbandment of the Niagara settlement was not about possession of land so much as it was about this access. The Neutral's permanent settlements at Niagara Falls were abandoned for several decades.[67]

Hennepin—writing in the late 1670s—reported that according to the Senecas, some Iroquois had moved away from their site at the falls. He said that besides the poisonous snakes, they had moved from "those thunders, those roarings" that came from under the falls "from fear of losing their hearing."[68] The people he wrote of were probably not Senecas, as elsewhere he referred to them by name. The Iroquois he met at Niagara were probably an amalgam of Senecas and the as-yet not fully assimilated remains of the Neutrals, who were Iroquoian, but not of the Five Nations, and were also occupying the westernmost of the Seneca settlements. In a culture where hearing signified life itself, the loss of hearing due to the thunders may have represented the loss of life to the warfare about to be set loose in the area.

The Neutrals who assimilated into the Senecas constructed their new identities as refugees from a great snake and disease, rather than as losers to the Senecas. It is only once the seventeenth-century outlines of Iroquoian soundways are known that the nineteenth-century tale can cast a light on the cultural process of mourning, warfare, and adoption from both the perspective of the Senecas and from the adopted survivors.

The Great Man of Power: African American Natural Soundways

James Albert Ukawsaw Gronniosaw, an African from Barnou, a thousand miles inland from the Gold Coast, wrote a deeply religious autobiography that was published in 1774. One of the turning points in his spiritual life took place in West Africa, probably in the 1740s, as he returned home from worshiping his West African gods with his family. A sudden storm arose, with "very heavy rain and thunder, more dreadful than ever I had heard: The heavens roared, and the earth trembled at it: I was highly affected and cast down; insomuch that I wept sadly, and could not follow my relations & friends home." He was overcome with fear. Later, he asked his mother "pray tell me who is the GREAT MAN of POWER that makes the thunder?" She replied that "there was no power but the sun, moon and stars; that they made all our country."[69]

Gronniosaw was writing in hindsight from the perspective of a converted Christian who was still enslaved, so his youthful concerns about the nature of thunder have to be read through the knowledge of what he would eventually become when he wrote about them. What he was hinting at in the passage about the sound of the thunder changing his life was that inside he was a Christian rather than a heathen, though at the time of the thunder he knew not what either was. Christians knew thunder to be the voice of God, the "GREAT MAN of POWER," as he so emphatically wrote it. The people of Barnou, in contrast, heard the thunder as a part of nature. The other world was inhabited only by a long string of ancestors, with no overarching deity. The stars and sun had created them long ago, and Gronniosaw's parents grew annoyed and worried with the boy's strange questions.

African Americans often treated thunder and lightning as an immanent willful force. The emphasis is often on lightning or thunder and lightning rather than thunder alone, although it is not clear whether this reflects a more visual culture among Africans or the attenuated soundways of the eighteenth-century European American amanuenses who wrote them down. In the eighteenth century, Phillis, an enslaved African American from Connecticut, asked to be buried under a tree when she died, "where the lightning [will] never find me." William D. Piersen maintains that this may have reflected a combination of a guilty conscience and the belief that lightning would seek out guilty people. African children in New England were taught that thunder was God's voice, but they often believed it was the voice of other beings as well. Cotton Mather thought that Africans were prone to devil worship, which was why they latched onto the fear of thunder so readily. These African American soundways corresponded with a widespread West African belief that thunder and lightning seek out and strike evildoers.[70]

West Africans had other ways of understanding thunder as an immanent and judgmental force. The best-known example of this is the story of Shango, who was at one time an extraordinarily powerful king of Yoruba. Shango was human, prone to outbursts of temper, and a perennial womanizer. After a fight among a number of his wives, the Yoruban myth claims, Shango stormed off in a fit of anger. All his wives then banded together to find him, cooperating in the process. When they found him, he refused to come back, saying that he would cause the women to start fighting again, so he ascended into the sky, where he became the master of thunder and lightning, striking down people who had cheated or done evil. Yoruban thunder and lightning represented raw power, and Shango was the dreadful and volatile avenger responsible for allocating it. Other natural sounds could also play important roles in African ideas of statecraft. In 1700, Capuchin missionaries in the Kingdom of Kongo reported that the King was not allowed to go to war without what Wyatt MacGaffey has described as "a favorable omen, the sound of the voices of the *bisimbi* as heard in the roaring waters of the river Ambriz."[71]

In planning the New York slave uprising of 1741, African American conspirators swore by thunder and lightning that they would not betray the plot.[72] They no doubt assumed that those who broke their vow ran the risk of angering the thunder. There may have been more to their swearing by thunder and lightning, though. The participants in the conspiracy were planning to take the law into their own hands in the face of what they considered to be an illegitimate rule that oppressed and enslaved them. In addition to calling for the thunder as simple witness to their vow, the conspirators may have been fashioning a legitimizing ritual out of the common denominator of the West African soundways that they knew. In Connecticut, eighteenth-century African Americans inaugurated their honorary governor by invoking the power of the thunder. In this case, they explicitly referred to thunder and lightning as a mark of legitimacy for their elected "officials."[73]

The New York and Connecticut examples parallel a founder's tale of the eighteenth-century Ashanti nation, located in what is now Ghana. The Ashanti tale related how their leader, Osei Tutu, unified the people into a single nation by warfare and alliances at the outset of the eighteenth century. Some time around 1700, Tutu's advisor, Anokye, devised the myth that a golden stool had descended from the sky in a great roar of thunder and landed in Tutu's lap. As it did in the story of Shango, thunder signified a medium for delivering great power, enough to construct, hold together, and maintain a state. The stool was a symbol that was equivalent to a European ruler's throne or scepter. According to Anokye's tale, the thunder literally brought the soul of the nation into Tutu's hands. These tales, much like the written constitutions of times yet to come, were the source of the legitimacy of Tutu's government. They explained from where the state derived its right to rule. In Tutu's case, the tale mapped out a sort of "divine right" theory of authority. This idea of a legitimizing ritual loses some of its exoticism if one considers the European rituals for taking possession of territory in the name of a particular ruler by planting crosses, getting papers signed, and firing off guns.[74] The underlying need to legitimize the exertion of force might be a generic or a universal. In contrast, the ways of satisfying that need—in the example above, by the animate roar of a great thunderclap—were culturally specific and significant.

African American soundways differed substantially from European American and Native American soundways. While the notion of Africans as primitive served an ideological purpose for whites, it also opened up a space in which African Americans were able to create a certain amount of autonomy. West African soundways that colonists could interpret as ethnic or "heathenish" often held multiple meanings. So when colonial pundits humorously noted that a black "governor" was sworn in using an invocation of thunder and lightning, they missed the ad hoc element of West African statecraft that claimed legitimacy for the black governor. Africans used these interstices and disjunctures between European American and African American soundways to create free spaces for themselves, even in the most unfree circumstances.

Colonizing the Natural Soundscape

European Americans, Native Americans, and African Americans each had distinctive soundways in regard to the natural world. They shared in common the drive to communicate with an animate and intelligent nature. But European American soundways shifted in the eighteenth century, so that many of the features of the natural soundscapes of the seventeenth century seem either alien or superstitious on first encounter. These older soundways did not disappear. As whiteness came to be the locus of power in the eighteenth century, the older beliefs were projected upon Native Americans and African Americans and marked as "primitive." Primitive, then, was not based on empirical observation so much as it was simply assumed. If the "primitives" were like an older form of European, it meant that European Americans not only could, but should slowly introduce them to civilization, or as it is now called, civil society. This ideology helped legitimate the colonization and domination of Native Americans and African Americans: it was for their own long-term good.[75]

The ideology was only complete from the standpoint of European Americans. On the one hand, Native Americans persisted with their own natural soundways, which show continuities from the seventeenth-century Jesuit accounts to the nineteenth-century Bureau of Ethnology accounts to modern folktales, ethnographies, and current practices. There was change over time, but there were also strong continuities that flew in the face of the civil/savage discourse. On the other hand, African Americans used the disjuncture between their attributed "primitive" nature and their own natural soundways to employ their own legitimizing rituals, unperceived by whites even as they recorded them. Both Native Americans and African Americans created and maintained natural soundways outside the domain of white colonizing discourses about natural sounds.

The distinction between civil and savage is relevant to another pair of binary opposites. It mirrors many of the traits of the literacy/orality distinction. Seen this way, the project of constructing ahistorical "oral" peoples with certain inherent traits can be understood as a modern legitimization of colonialism, whether internal, as in the case of Native Americans and African Americans by the United States, or external, as in the case of "western" cultures vis-à-vis today's "Other," often located in Africa or some part of the third world. While it is no longer proper to talk of them as primitive, or of westerners as especially civilized, it is still acceptable to speak of "oral" cultures and their failure or inability to embrace civil society.

Ultimately the task here is to historicize those categorized as "oral" and to show how that category holds better for a certain sector of literate seventeenth-century European Americans than for anyone else at that time. To do this, we must introduce the sounds that humans made: but how to do this when sounds are so evanescent? Natural sounds stayed more or less constant over time: thunder now sounds much as it did then. Perhaps some human sounds, then, are less ephemeral than scholars of orality can see as well. To begin to recover human-made sounds, we must postpone an engagement with the parts of the soundscape reducible to print and writing, and put off engaging with voice at all for the moment, as we turn to the instrumental sounds that people used to construct early American societies. There we will find soundscapes that can sometimes still be heard today.

CHAPTER TWO

From the Sounds of Things

"You'll think I'm loopy, but I tell you that bell was alive."
—Nobby Cranton, in Dorothy Sayers's *The Nine Tailors*

Sunday morning was quiet in downtown Philadelphia's historic district, a silence punctuated lightly by the occasional car whooshing past. As I walked the quarter mile from my parking spot to Christchurch, the eighteenth-century city's renowned Anglican house of worship, I was surprised to hear its bells slowly emerge from the stillness — not surprised to hear them, but surprised they were so quiet as to be barely audible a few blocks away. This was a ring of ten bells, eight of which were cast in 1749 at Whitechapel, England's leading bell foundry. The largest of them weighed 2,040 pounds, roughly the same as its broken but more famous Whitechapel sibling, the Liberty Bell. For a time they were the largest and loudest bells in the Americas. They were heard by shipmasters halfway across the Delaware River, a mile away. Their relative silence vexed me. The street noise that morning was if anything quieter than the hustle and bustle of eighteenth-century Philadelphia, with its iron-shod horses pulling iron-wheeled buggies across cobblestone streets during a time when the neighborhood served as a vital open-air center of the city instead of as a somewhat mummified tourist destination for secularly reverent citizens.

The carillonneur, Douglas Gefvert, who rings and takes care of the bells, provided the answer to the puzzle of the quiet bells after the first ringing died away. We had climbed rickety ladders up to the belfry. In a far corner hangs the small minister's bell, said to date from the 1690s, with a long rope dangling to the floor below. This is the only bell still rung the old way, by swinging the bell rather than just pulling the clapper. The other bells are played through a keyboard attached to ropes that swing the clappers into the sides of their respective bells. The bells themselves never move. The sound of the ringing is directed straight down

Fig. 2.1 Minister's bell, Christ Church, Philadelphia. Photograph by author.

into the floor of the belfry before it escapes through the belfry walls' sound openings, which begin a few feet above the floor, about halfway up the bells' sides. There are two reasons why the bells are quieter nowadays. . . . First, the belfry serves to muffle rather than amplify the ringing because the bells are no longer swung; thus the lips of the bells no longer point out the belfry openings when the bell sounds. Second, the swinging of the clapper rather than of the bell creates far fewer vibrations and thus much less volume. The trappings of the old way of ringing the bells remain, however, making it easier to explain not only why the bells are quieter now, but also why they were louder then.

Bells were usually hung from an axle affixed to a pulley having a radius a little more than the height of the bell. The ends of the axle rotated freely but were clamped down or set into a square heavy timber frame. Weighing hundreds of pounds, a bell was rung by pulling a rope attached to the pulley. The pulley used the bell's weight to help bring the bell nearly to an upside-down position, with the clapper held against the lowest part of the bell's sound bow (just above the lip) by inertia. A "stay" prevented the bell from turning all the way around in its frame. As the bell began its descent, the clapper was still ascending. When the highest part of the lip of the bell

Fig. 2.2 A bell hung in its frame. Adapted by author from Henry Thomas Ellacombe, *Practical Remarks of Belfries and Ringers* (London, 1859), Plate 6.

had descended about a third of the way, the clapper would strike the sound bow on the high side and project a note downward and outward off the lip of the bell. That is why bells are said to "ring out" when rung properly. The force of the descent would carry the bell, with an assisting pull from the bell ringer, to its ascent on the opposite side, the clapper resting against the sound bow again, repeating the process. A heavy bell took time to be rung, for its own weight had to be used to get it swinging properly. Bells sounded best when set high in a belfry designed to reverberate and amplify the sound. This kind of bell ringing put a tremendous stress on the structure that housed it. The frame that held the bells would be set diagonally rather than squared with the belfry walls, displacing the stress onto the strongest parts of the structure.

In the previous chapter we questioned a seemingly natural division in soundscapes, between what we now understand as agentless natural sounds and those that are made with human intent. In doing so, we uncovered beliefs about "natural" sounds that differed from more modern beliefs in two important ways, namely that all sounds had agents and that sounds had a tangible power that we no longer grant them. It is now time to return to sounds made by human agents, still leaving aside voice. These instrumental soundways all have earthly agents at the source, so the first feature of seventeenth-century natural soundscapes becomes moot. The second still holds. Instrumental sounds were granted a tangible power.

What can we tell about early America from the sounds of things? In this chapter, early American soundways are situated within the study of material life, the culture of things.[1] Although the instruments are inanimate, human minds and hands always crafted and used them. Moving from "natural" to human sounds, this chapter and the next consider instrumental soundways. Socially important instruments such as bells and drums are the part of the instrumental soundscape considered here. These instrumental sounds acted as a powerfully cohesive force used to build social order and govern traffic with worlds both visible and invisible.

Early Americans of all ethnicities carefully considered, designed, and used sonic instruments to create, challenge, and negotiate social and cultural milieus. First, European Americans used instrumental sounds to extend the limits of community beyond the realm of face-to-face encounters, and to shape social structures within or against those limits. In cross-cultural negotiations, European Americans encountered Native American instrumental soundways in which sounds stood in an important relation to group identity. African Americans used drums and other instruments to reconnect, often in new ways, their disrupted social, religious, and (where possible) political lives.

Focusing on instrumental soundways situates this part of the inquiry in between the nonhuman soundscape and orality. Like the study of "natural" soundways, it keeps us disentangled for a moment more from the thicket of theoretical issues surrounding the consideration of oral culture, and the latter's tendency to reduce the soundscape to spoken language. Instrumental sounds involve human manipulations of a soundscape, allowing us to consider some issues of human intent excluded from the preceding exploration of nonhuman sounds. In short, we can find out what people did with sound.

Attention to instrumental soundways underscores how mistaken it is to mentally separate church from state, sound from vision, motherland from colony, and heaven from earth in the seventeenth-century world. Michel Foucault has argued that these (and other) divisions were a mark of the Enlightenment, which replaced an earlier discourse (prevalent during the sixteenth and early seventeenth centuries). The earlier discourse, according to Foucault, emphasized relations, connections, and similarities instead of the Enlightenment attention to elements, taxonomies, and differences. In such worlds, the sounds of things tell us much because of sound's quality of mixing together rather than separating out.[2]

Bells, drums, and other instruments were all used to mediate between smaller social structures and larger identities (based in religious beliefs, town, region, nation, and colonial relations). Reconstructing the instrumental aspects of early American soundscapes requires a sharp modification of the thesis that the "literate" world of vision and the "oral" world of soundways resided on opposite sides of the divide between the "primitive" and the "civilized." A standard claim about so-called oral cultures is that they were worlds governed by face-to-face encounters. This visible limit to governance—no one could be part of a society without being visible in it— is treated as a hard boundary. Pre-modern communities could not grow without introducing some form of visible language that could extend the realm of the face-to-face. This notion of extension is key to distinguishing "civil" from "savage"—or "civilized" from "primitive." Rendering language visible—first in writing, then in print—is posited as what allowed for the extension of encounters beyond face-to-face exchange. Writing and print supposedly removed the necessity for the physical *presence* of an author— or authority—as a criterion for representation to take place.[3] However, that presence had never necessarily been *visible*, in the sense denoted by "face-to-face." Teleologically peering into the past (and not listening), present-day scholars have often implied that the visible world exhausts the field of possibilities in structuring social life. Sound has only counted when subsumed under vision. The "oral" in "oral culture" is thus face-to-face; it is generally that part of sounded language reducible to vision via writing and print. Subsumed in this way, it can seductively stand for the primitive past, a foil for civility, a non-extensive way of thinking.

The sonic means of extending civil society's reach that early Americans practiced, many of them not easily rendered visible, have been silenced in written and printed texts. While this has been the convention, there is no inherent reason that sonic evidence cannot be assessed as well as the visual. Soundways belong to a world set aside rather than lost. The material culture of soundways is much more permanent than scholars of orality would allow sound to be. Old rings of bells, for example, produce sound in the same ways now as then. Such things provide a record as useful as a text. Changes over time in their design, importance, and uses provide us with a means of better understanding early Americans' mental worlds.

Tintinnabulum Coelestis Benedictione Persunde

Europeans brought to the Americas a rich repertoire of material responses to a nonhuman sound, thunder. The practice of ringing bells to counter thunder's damaging power is the simplest, yet in some ways most difficult, point of departure for listening in on material soundways. It is simple because the sounds of thunder are more or less the same now as they were then. It is difficult because the belief that invisible agents can be reached via bells has been marginalized, giving way to observability as the chief criterion of truth. The beliefs no longer seem believable. They seem like magic, not science.

Since the Middle Ages, and perhaps before, Europeans believed that the sounds of great bells warded off thunder and lightning. For a church's bells to be struck and melted in their steeple, as they were in the English hamlet of Blechingley in 1607, was considered an ominous event. Simon Harward wrote *Discourse of Lightnings* because a storm had struck a neighboring town's bell tower, "melting into infinite fragments a goodly ring of bells."[4] Harward's readers wanted to know why. Bells were supposed to dispel the thunder, not thunder the bells. The explanations of how they were supposed to do this fell out along much the same lines as the explanation of thunder, into mechanical and spiritual forces, neither of which was exclusive of the other.

Catholics—and sometimes Anglicans—baptized church bells. Priests and congregants had long believed that thunder and lightning would destroy church bells that had not been "christened and hallowed." The ceremony did more than protect the bells. Late medieval Christians thought that a bell's baptism made it capable of dispelling thunderclouds. Once baptized, bells' inscriptions, saying things like *"fulgura flango"* (I subdue the thunderbolt), were thought to help as well. Ringing the bell carried the inscription to the heavens. Bell baptism had fallen into official disfavor during Elizabeth's reign, but was still widely believed to be effective when Harward wrote about Blechingley's ring of bells in 1607. Harward, whose beliefs leaned strongly toward the Reformation, pointed out that all but the newest of Blechingley's ancient church bells "had the blessing and baptizing at that time used and were hallooed by that prayer in the Masse booke," which read:

> *Almightie everlasting God, besprinkle this bell with thy heavenly blessing, that at the sound thereof, the fiery darts of the enemie, the stroke of thunderbolts, and hurts of tempests may farre be put to flight.*[5]

Upon first inquiry, particularly from the perspective of a *modern* literate person, Separatists and Dissenters appeared to consider bell baptisms as superstitious magic. It was under their pressure that the more conservative mainstream Anglican church officially stopped the practice. But the belief that all sounds had some willful being as their source, whether visible or invisible, was still quite alive at the outset of the seventeenth century. Bell baptisms even returned for a time during Archbishop Laud's conservative reform of the Church of England.[6]

Keith Thomas calls bell baptisms a form of "word magic," in which the bell was supposed to do its work by carrying the priest's invocation and the words inscribed in the bell itself to the heavens in its peals, there battling with the demons thought to be at the source of thunder, or else convincing God to take pity. But word magic only partially describes these English soundways. The Protestant rejection of word magic did not entail a disbelief in the efficacy of sound. English Protestants objected to the consecration of bells because word magic placed the priest or sorcerer above God, from whom they believed the thunder actually ensued. They even kept the idea of it as God's voice. Bell baptisms were frowned upon because they amounted to a contract that bound God. Attention to the full range of English soundways rather than just the orality allowed by the concept of word magic discloses a situation more complex than a simple decline in irrational superstitions.

In the early seventeenth century, the science that Thomas asserts killed word magic was more concerned with explaining the efficacy of willful sounds than with dismissing them as irrational. Scientific folk in early-seventeenth-century England believed "that by the stirring of the air" with the sounds of bells, "the cloudes may soon be dispersed or driven away." Here, sounds were conceived as tangible particles, part of a pre-Cartesian mechanical universe, spreading out like the ripples in a pond or a shotgun blast until they met and countered the sounds of the thunderstorm. Like countered like, as the English hoped that the powerful sounds of bells would mix with and attenuate the powerfully harmful sound of the thunder. "To shoot up ordinance into the aire," claimed Harward, was as effective as ringing a bell. The actual bullets and cannonballs had no effect, however. Like "our sight," such projectiles traveled in a "right line," displacing little of the vaporous matter constituting storm clouds. It was the sound of the shot that did the work. This, thought Harward, was also the reason that lightning was seen before thunder was heard. The former traveled directly, while the latter spread out, covering more territory, but slower.[7]

Reformed bells could neither bend God's will nor talk, but their sounds were still thought to be powerful in a tangible way in the seventeenth century. This scientific belief gave way only slowly, and for reformers and traditionalists alike, thunder was the voice of God, and the bells were the tiny sonic plea of the faithful for mercy from the power of that sound.

The Politics of Joy

Bells and other devices—some seldom thought of as sonic instruments—did more than ring out to the heavens; they rang in the state. According to David Cressy, the pan-European "vocabulary of celebration" included "ringing bells, shooting guns, sounding instruments, or raising cheers." England, and later British North America, employed this vocabulary in a distinctive way that harnessed these powerful sounds "to the needs of the state," even ringing it into being. "Joyful noises were made for the health of the King or Queen, for deliverance from the papacy, for victory and even defeat in battles." Cressy notes that "public celebration entwined the drives of *communitas* with the needs of power."[8] Societally sanctioned gunshot, cannon fire, drumming, song, bells, and other instruments all marked the emergence and development of national culture in its local instantiations.

Bell ringing and public noisemaking in general connected folk to community and community to the imagined nation and the invisible realm of the spirit. It is ill-advised to take apart too quickly the realms of church and state, of public and private, of visible and invisible, or of science, religion, and superstition, in a place where the ears of heaven could be bent if only the right channels were used, where the voice of God could be heard in a storm, and where ringing bells realized the will of the queen.

Instrumental sounds served local communities as much as they did an incipient British nationalism. People rang in their own communities in calling curfews—a practice dating back to William the Conqueror. They rang out on Rogation Days, when a priest would walk the perimeter of his parish ringing bells. With the very social order at stake, governing access to a town's bells was critically important. Bells protected the community and brought it together in tolling alarms for fires and calls to arms. In 1381, English bells were commandeered to start a peasant revolt. Rumors of a repeat abounded in Essex in 1566. In 1569, the North of England used bells to call its inhabitants to rebel.[9]

By the end of the sixteenth century, national unity began to be expressed by ringing bells on "crownation day" throughout all of England simultaneously. Across the country, the sounds of local bells marked the full extension of the nation. When they sounded a national identity into being by giving it a public hearing, their reasons were local and their own as well as national. They rang in the nation much like they rang in the years and the seasons.[10] It was a way of belonging to something larger than the face-to-face community, whether that something was located in space, as was the nation, or in time, as was the calendar and the seasons.

What happened to these instrumental soundways when they were carried over to places with no English institutional structures in place? Among the first things they did, colonial settlers put their soundscapes in order to create new societies. Besides recreating the familiar, the settlers used instrumental sounds to cross cultural boundaries and communicate in new ways.

Patience and Delivery

The 1609 hurricane that wrecked the *Sea Venture* deposited its passengers into a political theorist's dream laboratory, the uninhabited islands of Bermuda. The would-be Virginians had to build what civil and social order they could from scratch. Their ship was destroyed, leaving no hope for an immediate escape. Survival was not the issue. There were plenty of life-sustaining plants, a sea full of cattle-sized tortoises, fish, and other edibles, European pigs left on the island to breed decades before, and myriad birds.[11] But how would they create and maintain a social order? Admiral Somers took charge of trying to fashion new vessels. The crew salvaged parts from the wrecked *Sea Venture*. They scrapped all the wood, battered and cracked as it was. There was plenty of that available on the islands. Every scrap of metal, however, was carefully pulled out in the hope that they could fit new vessels with the salvage.

They put one metal instrument right to use rather than saving it for the new vessels. Gates, the presumed governor, used the *Sea Venture*'s bell to create and maintain a social order recognizable to the castaways. Strachey reported that "every morning and evening at the ringing of a bell" the whole company was gathered together, public prayers were said, and the roll called. Anyone not brought in by the bell was "duly punished." On Sundays, they were called by the bell twice more than usual for sermons on the importance of "thankfulness and unity, etc."[12] The bell served civil and religious purposes together, showing how closely the earthly and invisible worlds were intertwined in the early seventeenth century. It notified the castaways that it was time for both roll call and worship. It apprised the invisible world—and not just God, for this was reputedly the Devil's Island— that the congregation was assembling. Here was the adhesive Gates needed, and it held together not just the visible, but the invisible bonds that made for a working social order in the seventeenth century. The sound of the bell was powerful because it allowed Europeans to traffic in the invisible world as well as the visible. Rather than understanding the colonists as constructing some sort of face-to-face social order, we might better think of them as fluent in a mode of invisible sonic representation that was to be largely set aside for the visible world over the next two centuries. Through this aural mode, colonists took a step toward disembodying communication. This process is usually thought of as beginning with the telegraph and with radio broadcasting.[13] It would be mistaken, however, to think of these older, extensive instrumental sounds only in terms of future developments. Societally sanctioned instrumental sounds on Bermuda and elsewhere in the colonial Atlantic world made real bodies into incorporeal—but very real—entities. Incorporation is literally the making of such an imagined body.

In a land with no churches or courts, the sound of the bell served as the base on which to build social order. People had to wander off, out of sight of one another, in order to obtain food and other materials needed for rebuilding the ship. The bell called all within earshot together, literally ringing them in. Those beyond the bell's range, either beyond earshot or beyond obeying its toll, were in the wild. The would-be leaders feared that prolonged contact with wildness would tear the community apart, draining resources and threatening its survival.

Maintaining unity was a problem. From the start, some castaways had questioned Admiral Somers's authority on land. And while Thomas Gates had papers to govern Virginia, he had none for the Bermudas. Loss of manpower to factions and bickering would have set a dangerous example in a situation where the escape of any from the island depended on everyone working together. But some of the *Sea Venture*'s Company thought the island held out better prospects than Virginia, and tried to secede. A minister's clerk named Stephen Hopkins made "substantial arguments both civil and divine (scripture falsely quoted)" that Gates's authority as governor ceased with the shipwreck, and that "they were all then freed from the government of any man." For Gates to govern at all, such wildness had to be quashed. Gates used the *Sea Venture*'s bell to hold the provisional Bermudian society together. He had it rung to assemble the whole company, bringing Hopkins before them all in manacles. The chains bound the wildness and anarchy promised by Hopkins's words, but the sound of the bell was what bound the community, Hopkins, and Gates together in a public enactment of civil government. Hopkins was charged with mutiny and rebellion and sentenced to death, though he "made such a moan" that he was reprieved.[14]

During the nine months that they were stranded on the Bermudas, the sound of the wrecked ship's bell brought a sense of order and familiarity to what would otherwise have been a stateless chaos. Although Somers lost his life in the process, the castaways were able to build two new ships—the *Patience* and the *Delivery*—from local cedars and the salvaged rigging from the shipwreck. Gates maintained his government over the whole company for the duration of their stay on the island. On May 10, 1610, the survivors boarded the two new ships, no doubt ringing the bell once more to do so, and set sail for Virginia, weathering another tremendous thunderstorm off the coast of North America.[15]

Sounding the Chesapeake

After miraculously surviving a destructive natural storm, Gates landed in a civil tempest at the struggling colony of Jamestown. "Much grieved" by the "misery and misgovernment" of the colony, he set about restoring order as soon as his feet hit land. "First visiting the church," according to Strachey, Gates "caused the bell to be rung, at which all such as were able to come forth of their houses repaired to church." Those thus rung in heard a prayer after which Gates's commission was published by reading it aloud, installing him as governor by doing so. Gates's first three actions used sound—bells, preaching, and proclamation—to ritually reconstruct and redefine Jamestown, setting the people in their proper relation to God, to England, to Virginia's wilderness, and to each other. While the content of that restructuring may have been spelled out in the documents he carried, the act of setting it into motion was done with the chapel bell.

Soon after his being sworn in as governor, Gates installed a second bell

Fig. 2.3 Conjectural view of Jamestown, ca. 1614. Sometimes what is left out of a nontextual source can tell us as much as what has been included. The entire town was set up to be within earshot of the center, either to the sound of guns fired in warning from the periphery or from bells rung or guns shot from the center. The bells—there were two by 1614—were located at the traditional English location, the west end of the chapel (the left side in the drawing). The chapel is the largest building in the upper center of the enclosed fort compound. Perhaps the artist did not know the bells were there. Perhaps they were left out as unnecessary clutter and detail. They were important to the colony, though. Undated, unsigned drawing by Sidney E. King for the National Park Service, Colonial National Historic Park, Yorktown, Virginia.

at the west end of the chapel, perhaps to underscore the new order. The two bells mustered the troops as well as the faithful, and were used when colonists were punished for civil infractions as well as to serve notice to the heavens. Although no mention is made of using them to chase off thunderstorms or the plague, such practices were probably not too far removed from Gates's dispelling the bad air that hung over Jamestown with his bells.

Gates's peal sounded the limits of the colony's civil society. Those limits were influenced to a large degree by what was within earshot, a word that first appears in print in 1607.[16] Earshot was an effective measure of the limits of a community for two reasons. First, it extended the community beyond the face-to-face, as discussed above. Second, sounds intrude on the ears involuntarily. The sound of the bell was itself a force, a shot, not a declaration or a command. It compulsorily drew in all "who were able."

While bells held a particularly rich network of cultural meaning for the Jamestown voyagers, they were by no means the only instruments of sound that played an important role in ordering the society. High-status deaths, whether of people or the colony itself, were marked by volleys of gunshot and ordnance as well as bells.

Rituals of capture and possession had important sonic dimensions. When the leaders of the first Jamestown expedition agreed to a site on which to plant the colony in 1607, they let loose a ritualized fanfare of trumpets as part of the process of legitimating their claim upon the land for colony and king. Along with drums their sound was used to hail enemy ships in battle as well as to signal their own ships in peace. When Jamestown was briefly abandoned in 1610, Gates "commanded every man at the beating of the drum to repair aboard" the departing ships. They left "about noon, giving a farewell with a peal of small shot" to an invisible audience. A few hours later they serendipitously ran into a new supply convoy carrying Lord De La Warre to Jamestown, so they turned around and went back. When De La Warre arrived, the settlers were assembled by the sound of the bell and he became governor with the public reading aloud of his commission.

Instrumental sounds were also valued at sea. Trumpeters and drummers were a regular component of ships' crews. Trumpeters were usually thought important enough to garner a quadruple share of pay, the same as a boatswain, or a surgeon, and almost double that of common sailors. To capture an enemy ship, one had to "sound drums and trumpets and St. George for England" (and of course, win the battle). Once possessed, "out goes the boat, they are launched from the ship side, entertaine them with a generall cry, God save the Captaine and all the company with the trumpets sounding." More mundane tasks also required trumpet blasts to be carried out properly. Smith advised would-be shipmasters:

> The Trumpeter is always to attend the captains command, and to sound either at his going ashore or comming aboord, at the entertainment of strangers, also when you hale a ship, when you charge board or enter; and the poope is his place to stand or sit upon, if there be a noise they are to attend to him, if there be not every one he doth teach to beare a part the Captaine is to incourage him, by increasing his shares, or pay and give the master Trumpeter a reward.[17]

Table 2.1. John Smith's Prescribed Rates of Pay for Shipboard Duties

	Accidence	Sea Grammar
Captain	9 shares	9–10 shares
Lieutenant	neg. w/ Capt.	9 or neg. w/ Capt.
Master	7	7–8
Mates	5	5–7
Chirurgion	3	3–6
Gunners	5	5–6
Botswain	4	5–6
Marshall	4	5–6[1]
Carpenter	5	5–6
Trumpeter	4	5–6
Quartermasters	4	4–5
Cooper	—	4–5
Chir.'s mate	—	4–5
Gunner's Mate	—	4–5
Carp.'s Mate	—	4–5
Corporal	3	3–4
Quarter Gunners	—	3–4
Trump.'s Mate	—	3 1/2–4[2]
Steward	3	3–4
Cook	3	3–4
Coxswaine	3	3–4
Swabber	—	3–4
Sailors[3]	1 1/2–2	1/2–3
Boys	1	1–1 1/2

Source: Smith, *Sea Grammar*, 2: 110–11, and Smith, *Accidence*, 2: 26–27.

[1] "On English ships they seldom use any Marshall, whose shares amongst the French are equall with the Botswaines," wrote Smith. The marshall was usually found on French ships and was responsible for carrying out punishments.

[2] "4" mistakenly printed as "3."

[3] Called "Younkers" in *Sea Grammar*

Ship's captains were not above recreational uses of trumpeters' skills either. Father Andrew White recounted an exciting race in the 1630s between the ship he was on and another named the *Dragon* for about an hour on the high seas with a good wind. As the ships raced neck and neck, the passengers and crew were treated to "the pleasant sound of trumpetts."[18]

Sounds were effective. The first generation of English Virginians were pragmatists about the sounds they chose to manipulate. When they could, they used all the sounds above for doing things. Sounds could be powerful, and powerful sounds were the ones that interested them. Loud sounds impressed them most. This makes sense when older theories of acoustics are considered: the louder the sound, the more force behind it, and the more that could be done with it. Bells, gunshot, trumpets, and drums: all were ways of making sounds louder, thus amplifying or extending the range of earshot. One consequence of this soundway was the ability to push the limits of community and civil order beyond face-to-face contact. Print also extended these capabilities using the visible world, but in different ways, with different consequences.

Sounds with no visible sources, such as thunder, were considered portentous for precisely that reason. So too with human sounds. Obviously, the immediate source behind the ringing of bells was human—except in the case of earthquakes, where the ringing of bells was portentous indeed. There could be invisible human sources as well. Thus Gates could cause the bell to ring without ever touching it and its peal could assemble the community because it carried not only Gates' will, but the powers of state that descended through the Virginia company's royal charter, which in turn came from the king, whose power came from divine right. All sounds had agents, and the more human the sound, the more agents it was likely to have accrued in being issued, particularly when that sound was an integral part of defining the limits of a civil society where church and state functioned together.[19]

Conversing with Powhatan

So far we have been considering the colonists as if they were conversing only with themselves and the heavens. Jamestown's earliest settlers were very attentive to First Nations soundways, and vice versa. At first, both Native Americans and the new settlers had the obstacle of language separating them, so the instrumental sounds they made often stood in for language, making them doubly significant. While for Europeans loud intercultural sounds were part of "rituals of possession," for Native Americans they were important markers of group identity. The sounds of gunshot were quickly associated with the comings and goings of the invaders, and became expected.

European Americans understood that Indians were impressed by loud sounds but often missed the association with group identity. Father White, a missionary to Maryland, reported in 1634 that the Chesapeake Americans "trembled to hear our ordinance, thinking them fearefuller than any thunder they had ever heard." John Smith maintained that Indians throughout the Chesapeake revered loud sounds such as thunder and the reports of the colonists' guns. The Powhatans had a concept similar to *manitou,* which Smith made the most of. At the Tockwough River, he gained Indian allies by "firing 2 or 3 rackets [rockets]" over the river. On account of making such a racket, Smith claimed, they "supposed nothing impossible wee attempted." When the colonists crowned Powhatan as a vassal of the king, the colony's boats fired off such a huge "volley of shot that the king start[ed] up in horrible feare" for a moment before regaining his composure.[20]

On more than one occasion, the sound of guns was enough to repel an attack. While exploring the upper Chesapeake, Smith and his small company were ambushed by over a hundred Potomac warriors. The colonists responded with gunfire, but not to hit anyone. Smith reported that "the grazing of the bullets upon the river, with the ecco of the woods[,] so amazed them" that they threw down their bows and arrows and, exchanging hostages as a gesture of good faith and for collateral, they all went together to the Potomacs' town, where the colonists were treated well. Other times, Smith announced his entrance to an Indian town by firing several shots in the air, claiming it ensured his safety.[21] Perhaps as much as scaring the Indians, he was reassuring himself, clearing the air.

When Smith went to barter with one group of Indians for food, they asked as part of the bargain to hear the party's guns. Smith and company fired them off in a riverbed, "which in regard of the eccho seemed a peale of ordnance." Smith knew it was the sound that impressed them as well as the bullets, so he used the acoustics of the riverbed to maximize the effect.[22]

Small sounds had their place, too. Many Chesapeake area natives trimmed their clothes with shells and snake rattles to "make a Certayne murmering or whistling noyse by gathereing the wind, in which they seeme to take great jollety, and [they] hold that a kind of bravery." They used the word bravery in its now obsolete sense of ostentatious finery or adornment, a thing of beauty or interest, something to exhibit, or that which is worthy of boasting. In the controversial passage of Smith's *Generall Historie* in which the young daughter of Powhatan, Pocahantas, prevented Smith's execution, Powhatan was supposedly "contented" that Smith should be spared to make hatchets for him and make "bells, beads, and copper" for Pocahontas. Smith mentions giving a few bells as payment to Indians whose labors would have cost him a horse in England.[23] Another time, the food obtained in exchange for bells may have kept the colony alive. The Powhatans' desire for new and unfamiliar sounds drove these exchanges.

Chesapeake Indians listened carefully, constructing specialized physical spaces for the spoken word and for the sounds of the world around them. They built scaffolds specifically for holding conversations. Unlike the public acoustical spaces that Europeans made for church oratory, Chesapeakes made their speaking platforms to underscore the importance of one-to-

Fig. 2.4 A listening post on the edge of a ripe cornfield. This print was made by Theodore DeBry from watercolor paintings by John White. Courtesy of the Library of Congress.

one conversations that Anglo-Americans would have held more privately. This hints at public and private being constructed along different lines in native and white communities. Theodore DeBry's engraving (based on eyewitness John White's watercolor) of a Secotan village shows listening posts in the American fields that reflected and amplified sounds.[24] These could be used to scare off birds, but their parabolic shape also collected sounds, making this an audible sentry post—especially useful in the presence of English men willing to pilfer corn to stay alive.

The significance of instrumental sounds to First Nations people was not lost on the colonists. "Wee might to this daye," argued Smith, "have wrought more amongst them by the Beating of a Drumme, that [than?] now wee can by the fieringe of a Canon." Powhatan's people were edgy about the coming of the English. Strachey wrote that "straunge whispers (indeed) and secrett[s] at this hower run among these people and possess them with amazement. . . . Every newes and blast of rumour strykes them. . . . The noyses of our drumms of our shrill Trumpets and great Ordinance terrefyes them so as they startle at the Report of them, how soever far from the reach of daunger." Having gone out on another food-bartering expedition, Smith, Captain Newport, and Master Scrivener left their boat and marched toward the sachem Powhatan with one of them blasting a trumpet for effect. Smith impressed Powhatan with descriptions of European wars featuring trumpets and drums. At another juncture when things were tense between the two leaders, Smith told Powhatan that the Indians will know when the English are going to fight because they will sound their "drums and trumpets."[25]

Bells, guns, and trumpets also alerted the colonists of impending danger. Once, while men were planting corn and cutting down trees outside the safety of the fort, they heard an alarm—either gunshot or bells—from the village. Thinking this warning from the center was an Indian attack, they were relieved to discover it was a long-overdue supply ship. Another time, Opechancanough captured Smith while he was split off from the other two members of his party. The two were supposed to fire their guns at the first sign of Indians to warn him. Hearing "a loud cry and a halloing of Indians, but no warning peece," Smith knew he was in trouble and that his companions were captured or dead. When Opechancanough's men surrounded Smith, he fired his gun a few times, the sound of which kept them back, but ultimately he slipped in some mud and had to surrender his guns.[26]

At the behest of the Indians as much as the colonists, political comings, goings, agreements, and wars were all publicly marked with great sounds. Upon departing from a particularly friendly diplomatic foray to the Indians on the eastern shore of the Chesapeake, the colonists—at the parting request of their hosts—fired off a loud volley of gunfire to which the Indians responded with a loud shout.[27] This exchange of sounds was a demonstration of political power and identity, mutually understood across barriers of language and culture. The terms of the exchange were set as much by the Indians as the colonists.

The colonists' lives depended on properly understanding their new soundscapes. Strachey noted that alarms traveled up the James River faster than the colonists ever could. Their comings and goings were always known in advance by the Indians. Communication networks along the shoreline were well established and quick. Smith, not content to observe, tested the speed of the network by spreading false rumors to Indians near Jamestown and then traveling upriver, where he would hear the same rumor repeated.[28] The idea of a communication network would have been odd to the colonists in this time before the advent of media that could travel faster than human messengers. The Indians' quick communication would have seemed impossible or even diabolical.

In the second decade of the seventeenth century, Powhatan and then Opechancanough lulled the colonists into a belief that all was well. They began to let down their guard, gradually drifting out of earshot from each other to start small plantations, further encroaching on Indian lands. On March 22, 1622, Opechancanough put the full force of the First Nations communication network into action. The Indians attacked the widely scattered settlements "at one instant," hitting plantations "one hundred and fortie miles up on [the James] River on both sides." Smith found it particularly remarkable that even though the Indian settlements were as small and scattered as the English ones, the Indians were able to act in concert. The Indians killed some 347 colonists that day, nearly a third of the English population. Although the argument is implicit rather than explicit, the Indians had communication networks that the colonists admired but did not understand. Smith was unable to explain how the widely scattered Indians had been in such good communication when they were not physically together. The colonists responded by once again ringing in their settlements. Twenty-five of the Virginia settlements were ordered abandoned, and the settlers moved into the remaining six. Lacking the Native Americans' skills at communicating effectively over a scattered area, the colonists returned to within earshot of their fellows for safety. The era of intercultural communication was over, much too late by Smith's estimate. The English cut off general communications with the Indians and sought to remove them from the Chesapeake altogether by any means necessary.[29]

Rung in by bells, never safely living beyond earshot—no matter how well extended by guns, trumpets, shouts, and bells—instrumental soundways and their consequent soundscapes bound the early Chesapeake colonists together, but not necessarily so close as a face-to-face community. The colonists pushed the Indians back in the 1620s and 1630s and began once again to spread out of earshot. They would suffer another attack in 1645, but by that time there were too many English invaders to wipe them out completely. The colonists were there to stay, and a new tobacco-based plantation economy began to take hold. Later seventeenth-century Chesapeake churches seldom had bells, because ringing them was a futile exercise in a plantation economy where everyone lived far apart from neighbors and town.[30] The plantation or farm became the locus of community rather than the town or village. It also became the new place for bells, used to order a different sonic regime than that negotiated between the Indians and the first generation of settlers. The first generation of colonists did not simply choose to believe in powerful sounds; they had no other set of beliefs by which to live.

Bells, Drums, and Shells in New England

As in early Jamestown, New England towns used instrumental sounds to order their worlds. Bells were important from the very beginning of Puritan New England, and great effort went into obtaining the best possible instruments. They were usually crafted in England, although some were booty from captured buccaneers and others were imported from the

Dutch.[31] Cambridge (known until 1638 as Newtown) had a bell installed at the top of its first meetinghouse in 1632. Hingham owned a bell by 1633, Salem by 1638, Newton (then known as Cambridge Village) by 1639, Boston by 1641, Woburn by 1642, Watertown by 1648, Springfield by 1653, Charlestown by 1657, Malden by 1658, Ipswich by 1659, Dorchester by 1662, Newbury by 1665, Hadley by 1670, and Hartford by 1665.[32]

While Cressy's assertion that "early American churches had no bells to ring" does not hold up to scrutiny, many New England towns used instruments other than bells to ring in their inhabitants. Dorchester used drums at least part of the time until 1662, but they also had a bell frame separate from their second meetinghouse (built in 1645), indicating a bell was in use there at least part of the time. By 1674 Dorchester's bell was "broken, and it may be, dangerous to be rung." It was taken down and sent to England to be replaced or repaired. By 1680 they had a new one, but the town gained a little infamy for its bell-less interludes in the popular *Ballad of New England*, which chided Dorchester:

> Well, that night I slept 'til near prayer time,
> Next morning I wondered I heard no bells chime:
> At which I did ask and the reason I found
> 'Twas because they had ne'er a bell in the town.

Cambridge replaced its bell with a drummer from 1636 to 1648. Smaller towns and frontier towns often did without, commonly blowing conch shells or using drums to summon the people together. Haverhill used a

Fig. 2.5 Seventeenth-century drum. Courtesy of the Connecticut Historical Society.

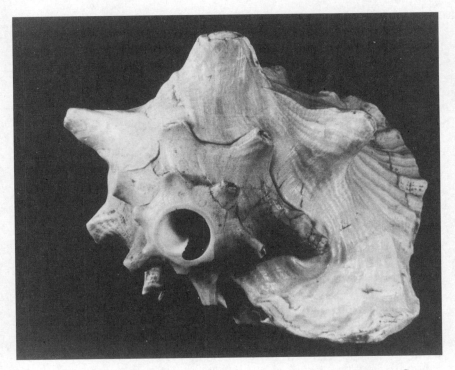

Fig. 2.6 Conch shell used in Whately, Connecticut, during the seventeenth century. Courtesy of the Dublin Seminar for New England Folklife.

drum or a horn to call people to church and to town meetings from its founding in 1640 until it imported a bell from London in 1748. Huntington, Long Island, had no bell at its church from the founding of the town in 1653 until 1715. During that time the militia company drummer called the faithful to worship. Greenfield and Northfield used drums. The towns of South Hadley, Montague, Shelburne, and Whately used conch shells.[33]

Whether bells, drums, or conch shells, these public instruments regulated many aspects of daily life in seventeenth-century New England. Newton kept its residents within the sound of its bell, passing a law in 1685 prohibiting anyone from living more than half a mile away. When Newbury residents sought a new bell in 1700, they instructed the buyer to obtain one loud enough for all the residents to be able to hear. Hartford selectmen made the connection between bells and social order explicit in 1665, ordering a curfew bell to be rung at nine in the evening "to prevent disorderly meetings" among the townspeople. Salem passed an ordinance in

1673 that the town's bell be rung every morning at five and every evening at nine from spring until fall "as an admonition to improve the light of day and keep good hours at night." Seventeenth-century Boston, Charlestown, Ipswich, Newbury, and Hadley all rang a curfew bell to regulate and order nightlife, too. John Josselyn, visiting Boston in 1663, reported that young men would court women by walking with them on the common "till the nine a clock bell rings them home to their respective habitations." Newbury bell ringers adopted the old English practice of tolling the day of the month each night after the curfew bell.[34]

The sound of the drums or bells could simultaneously be an enactment and a justification of authority. The hanging of three Quakers in Boston provides a good example. Boston officials ordered the town drummer to drown out the gallows speeches of Mary Dyer, William Robinson, and Marmaduke Stephenson. The three were hanged in 1659 because they refused to stop preaching in Massachusetts. Boston leaders had sentenced the Quakers to death for threatening civil and religious order with their speeches. The use of drums to drown out any further vocal threats emphasized the importance of sounds both mentally and practically. The speech of the Quakers was important enough to warrant a very loud silencing of their last words with drums, and a properly run civil and ecclesiastical order possessed the means of being louder.[35]

Bells sounded the cadence of everyday life in New England. Funeral and wedding bells knelled important passages in the lives of individuals. Meetinghouse bells summoned the faithful to church each Sunday. There was the pardon bell to forgive sinners, the Gabriel bell to awaken people from their slumber, and the pudding bell, tolled to let the cooks know that services had let out. The same bells served notice to the invisible world as well. Although Puritans may have rejected the notion that bells would drive off demons, the bells would let God as well as the godly know of services about to begin.[36]

As in Jamestown, church and state were inextricably intertwined. The bells (or drums or conch shells) that summoned churchgoers also "warned" them—now as citizens—to town meetings and militia days. Many towns employed a night watchman, like David Ray of Charlestown, who was paid to walk about "with his bell every night from eleven o'clock until five in the morning, to keep watch for alarums and fires, and give timely notice thereof." The watchman may have carried a handbell, like those rung by town criers when they published the news. His alarms were a sanctioned form of publication at a time when publication took place at least as often sonically as it did visually in print or manuscript.[37]

The spiritual world could warn back, too. Joseph Glanvill, a moderate

Source: St. George, _Conversing By Signs_ 187

Fig. 2.7 The demonic drummer of Tedworth beating a tattoo in the air. The illustrator attributed the visual characteristics to the sounds Glanvill heard. Courtesy of the Library of Congress.

Anglican and a member of the Royal Society, made a scientific defense of demons and witches in the seventeenth century that was influential in New England as well as in Britain. His first and most widely celebrated case was the "Drummer of Tedworth," a demon to whom he attributed the regular tapping and noises that occurred in the house of a Mr. Mompesson in Wilshire. Increase Mather reported that in 1679, William Morse's wife "heard a noise upon the roof of their House" in Newbury, Connecticut, "as if Sticks and Stones had been thrown against it with great violence." This turned out to be the diabolical opening tattoo that culminated in the possession of Morse's son. When Cotton Mather sought to defend the belief in witchcraft in the midst of the Salem outbreaks, he recounted the story of a town prone to Sabbath breaking, profanity, and drunkenness whose church rolls increased dramatically when one night "there was heard a Great Noise, with Rattling of Chains, up and down the Town," causing the guilty to fear that the devil was there to claim them.[38]

Bells marked New Englanders' transatlantic connections. They were perhaps the most expensive ornament of English identity. The sting of the *New England Ballad* for New Englanders was that they were deficient on the matter of bells rather than opposed to them in principle. Until the eighteenth century, nearly all of New England's bells came from old England. Town records often made note of the sea captain responsible for carrying the bell over. Occasionally bells had more colorful connections. The town of Beverly's first bell was seized in a Massachusetts raid on the Catholic friary at Port Royal in 1656. Sandwich's bell, cast at a Dutch foundry in 1675, came from the widow of a shipwrecked sea captain, perhaps salvaged from the wreck that took her husband's life.[39]

Some Puritans opposed bells on ideological grounds, but this was hardly the norm. The minister of Strawberry Bank (now Portsmouth, New Hampshire) argued to his congregation that a bell would be popish. He was successfully opposed by a sea captain named Jackson, who swung the vote by bribing some sailors to take his side. Jackson went to England and returned with a loud fog bell. The minister, discovering the bribe during Jackson's absence, brought the congregation around to his point of view and the bell sat on the wharf for the duration of the minister's tenure. Perhaps this was why Boston newspapers noted that lightning had damaged the meetinghouse in Portsmouth in 1736 and 1759, but had left the bells intact. Quaker George Keith made the most principled argument against bells, saying true believers did not need "an outward Bell, hanging in a Steeple, to call them together, but the Gospel-Bell did ring and sound in their hearts." Even this, however, did not prevent bells from being rung in Pennsylvania. The Massachusetts town of Hingham rejected a bell in 1635, saying their drum was good enough, but soon after they obtained a bell anyway. In 1685, they voted to procure a new English bell "as big againe as the old one was if it may be had" for their new meetinghouse. A sea captain named Hubbard brought them one that year, charging them five pounds and four shillings for it. Offsetting the claims that bells were "popish" was the fact that Puritans had not been allowed to proclaim their religious services publicly in England. Many felt that ringing a church bell or proclaiming a town meeting was an act of sonic identity that tied them to the heavens above and their English ways even while asserting their liberty from the anti-Puritanism they had fled.[40]

Some bells were too large for the belfry or turret that housed them and actually swayed the meetinghouse when rung. Often, church bells would

Fig. 2.8 An old free-standing belfry in Lexington, Massachusetts. Courtesy of the Library of Congress.

have to be removed from the belfry and housed outside in order to save the structure of the meetinghouse. The image of a bell rocking the church foundations underscores the degree to which Puritans valued the sounds of their bells. The selectmen of Newton ordered the bell to be removed from the top of the meetinghouse and rung from "some convenient place for the benefit of the town" rather than just removing it. Malden had trouble with its belfry and removed the bell to the top of a nearby rock to give it more support in 1684. A workman was commissioned to rebuild it six years later, but four years after that they still had troubles with the "tarat" (turret) for the bell. Dorchester, Medfield, and Concord removed their bells to separate structures, indicating they were too big for their belfries.[41]

While obtaining and maintaining a bell was costly, its sound was an immediate marker of authority and social order. Bell ringing, however, was a lowly occupation. The ringer was usually also hired as the janitor of the church and often took on other jobs like digging graves, carrying baptismal water to the meetinghouse, shutting the casements during storms, shoveling snow, and chasing dogs out of meetings. In Dorchester, the bell ringer's widow took over his job after he died, so it was not strictly a man's job.[42]

Native Americans interpreted instrumental sounds as markers of group identity. European Americans both extended and marked the limits of community with bells, guns, drums, and shells. As the seventeenth century proceeded, an increasingly significant number of enslaved Africans came to constitute the workforce, not only in the South but in northern and middle colony plantations and cities as well. They were all—enslaved and free alike—knit into a larger system, the American side of the Atlantic world. For North Americans, in a colonial backwater, the centers of vision were not just England, France, Holland, and Spain, but also Africa and the much more successful Caribbean plantations. Thus in the next section we will borrow from Caribbean and African sources to find out about the sounds of things African—here a simple page of sheet music and some drawings of homemade instruments.

An Evening's Entertainment

Before 1688, African slaves on southern Jamaican plantations had never seen anyone from the British Royal Society.[43] That year, Dr. Hans Sloane left his residence at "King's Hall"—the governor's estate in Spanish Town—to visit a sugar plantation in the interior. He was not only a Royal Society fellow but physician and friend to the governor of the island, Christopher, Duke of Abermarle. Sloane's purpose was to make field notes for his catalogue of New World plant and animal life.[44] For diversion, the plantation owner invited Sloane and another guest, a French musician named Baptiste, to witness a festival held by the slaves of the estate. To European senses, such music and revelry presented a spectacle not to be missed. Though the owner was uneasy because of recent revolts and unrest among the slaves of nearby plantations, he, his overseer, and his guests left the great house one evening and hiked a short distance down a path into a wooded area.

They entered a clearing that had been picked clean of grass and carefully swept. In the center, two African musicians sat on logs by a small fire of corn stubble, playing homemade string instruments. About a dozen African men and women with "Rattles ty'd to their Legs and Wrists, and in their Hands," were dancing and singing inside a ring of people surrounding the fire. The others clapped or scraped sticks, shook bean pods, or beat on an iron hoe blade. Together, these dancers and percussionists made "a noise, keeping time with one who makes a sound answering it on the mouth of an empty Gourd or Jar with his Hand."[45] After a short while, the revelers began to realize they had visitors, and the music gradually wound down and stopped. Baptiste, who had been taking notes, whispered something in French to Sloane, who was making rough sketches of the instruments that the two Africans were playing.

The overseer shouted at the slaves in a pidgin language, directing them to continue. The musicians resumed with a different song, this one with no words. Its simple repeated melody at first sounded dissonant to Sloane, but after a few minutes, he and his companions became entranced by the rolling rhythm of the dancers and clappers, the counterpoint of the gourd-beaters and the repetitious melody, all locked together. The time flew by unobserved until this piece also wound to its end.

Sloane then asked the master whether he could question the slaves about the music. The master, quite uneasy by that time, pointed out that it would be fruitless to try to communicate directly with them and called to his overseer, who explained a little about the slaves while a new piece of music began. A man playing an instrument unfamiliar to the Europeans accompanied a single singer in this quieter, more melodic song. While Sloane spoke with the overseer and the master, Baptiste, disregarding the English conversation, applied himself to notating the rhythm and melody of the music as best he could. Afterward he filled in the few fragments of lyric he could piece together from memory.

The music ended and the ring of people began to disperse. The overseer called out to one of the musicians and questioned him in pidgin. The language they used was suited to the one-way communication of work commands but was a source of confusion in this context. At length, the overseer turned to Sloane and told him what he believed to be the African origin of each song in turn: Angola, Papa, and Koromanti.

A remarkable scene: several languages—pidgin, English, French, at least

Upon one of their Feſtivals when a great many of the Negro Muſicians were gathered together, I deſired Mr. *Baptiſte,* the beſt Muſician there to take the Words they ſung and ſet them to Muſick, which follows.

You muſt clap Hands when the Baſe is plaid, and cry, *Alla, Alla.*

Fig. 2.9 Three African songs from Jamaica, 1688. The song titles refer to West and Central African ethnicities. Hans Sloane, *A Voyage to the Islands of Madera, Barbados, Nieves, St. Christopher, and Jamaica,* vol. 1 (London, 1707). Courtesy of the Trustees of the Boston Public Library.

two (and probably more) unrelated African tongues—three discrete musical styles being recorded by someone versed in a fourth, participants ranging from slaves to gentry, with connections to three continents, all thrown together for a moment in time.

Pidginization and Creolization

The music that Sloane and Baptiste recorded is fascinating in its own right; it is also important in a larger historical context. The music and descriptions yield insights into one of the fundamental problems challenging African American history. On one hand, some argue that slavery was so traumatic that it destroyed any usable African past. On the other are countless examples of persistent Africanisms in the Americas. A mediating position is now commonly agreed upon, positing creolization—a particular form of cultural mixing—as the solution to the debate. Often, however, the details of how the processes of transit and transition have actually played out remain vague.[46]

One problem that remains is that early American evidence of African culture formation is fragmentary at best. Kenneth Bilby, an anthropologist who has studied the cultural origins of the Jamaican Maroons, maintains that the sketchy nature of the historical literature generally precludes the recovery of the earliest workings of creolization among enslaved Africans, especially "at the individual level, where conscious creative decisions (as well as unconscious adjustments) were made." Bilby rightly asserts that although the beginnings of the process have been obscured, "the concrete results are nonetheless visible everywhere, both in the documented music and dance of the past and in the many continuities in context, style and form displayed by their present-day musical descendants."[47]

The sonic descriptions in Sloane's *Voyages* provide a unique glimpse into the workings of creolization among enslaved Africans of known ethnicity, and thus into the origins of African identities in the Americas. Baptiste unintentionally rendered a number of distinctly African features that were not yet recognized or employed in the seventeenth-century European music with which he would have been familiar.[48] Sloane supplies an impression not only of diverse first-generation musical soundways, but also of how enslaved Africans selected, combined, and changed these soundways to produce something new.

The linguistic paradigm of pidginization provides a way of drawing out the musical evidence in Sloane and connecting it to more general models of cultural transmission. From the perspective of the first generations of enslaved Africans in the Americas, pidginization was a process of experimenting, tentative and provisional, by which people from different cultures adapted to each other and new environments as best they could.[49] Planters had the power to create groups of dislocated peoples. Although they controlled the structures of these groups, they were unable and unwilling to control the cultural contents of the local communities that such groups built within the structures. Enslaved Africans used sound to create cultural identities within the contents of this space. Planters distanced themselves from the everyday activities of the enslaved, leaving local communities in relative isolation from English culture except as it related to everyday work activities. While the planter goal of mixing diverse ethnicities was not met to the degree it was prescribed, enslaved Africans found themselves in groups that most often had no culture or language native to all. Pidginization, of both languages and cultures, was the way that they negotiated viable communities in the first generation.[50]

Children born into slavery combined, stabilized, and expanded the pidginized, African, and English constituents of their cultural environment to produce a creole language and culture. They actively if unconsciously made sense of a fragmented world. This is the process of creolization, quite different from pidginization.

Music bears more than a surface relationship to language. A culture's music has a phonology of aesthetically permissible notes, a vocabulary of acceptable scales and rhythms, and a syntax of customs and rules that govern the largely unconscious ways people represent themselves through these notes, scales, and rhythms to produce what they recognize as music. These soundways are conditioned by the cultural community—in the case of settled cultures, by means of tradition and adaptation; in the case of displaced African ethnic cultures meeting in the bonds of slavery, by means of negotiation.[51]

Angola, Papa, and Koromanti

The three musical transcriptions in Sloane are headed "Angola," "Papa," and "Koromanti." He thought the labels named the origin of each piece. Between 1655 and 1680, one-fourth to one-third of Jamaican slave imports were from other islands in the West Indies. Many were Africans who had been seasoned in Barbados. The remainder came directly from Africa, mostly in ships of the Royal African Company that obtained their human cargos at ports between the Senegal and Niger rivers in West Africa. In addition, Jamaican planters illegally purchased a small number of Central Africans from freelance and Portuguese slavers. Jamaica planters used "Koromanti" and its variant spellings—"Coramantee," "Coromantyn," and "Kromanti," among others—for slaves from the Akan region of the West African Gold Coast (modern Ghana), after the port of Coromantin there. Koromantis spoke languages from the Western Kwa family, which was in turn a subdivision of the more ancient and loosely related Kwa language family that extended from what is now the Ivory Coast to Cameroon. The Koromantis were the most prized slaves in Jamaica during the earliest years of British settlement. They were also very troublesome. Many of them had strong military backgrounds, having fought in local wars throughout the seventeenth century among the littoral and forest states of the Gold Coast. The Jamaican landscape offered an ideal arena for their style of open-formation fighting; hence, they were able to mount and sustain rebellions between 1673 and 1686. Rebels who were not caught, mostly Koromantis, retreated to the mountains in the parishes of St. Ann, St. Elizabeth, and Clarendon to form the core of the Maroon communities, one of which, in St. Elizabeth Parish, persists to this day.[52]

Perhaps as a result of these revolts, planters purchased a greater number of Angolan slaves in the last quarter of the seventeenth century. In European parlance, "Angola" referred to a vast area of Central Africa that included Kongo, Angola, Loango, and a number of smaller states. All of these peoples spoke Bantu languages and shared many cultural traits, making the transition from one Central African culture to another a relatively easy process. From about 1680 to 1700, Angolans in this broad sense constituted about forty percent of all slaves imported to Jamaica. In practice, planters found them hard to control. They showed a decided tendency to run away, compounded by their belief that "on their deaths they are going home again," to Central Africa, which Sloane concluded was "no luciferous Experiment, for on hard usage they kill themselves." These factors led planters to return to Koromantis and related Gold Coast ethnicities by the end of the century.[53]

After the Koromanti-led revolts of the 1670s and 1680s, planters also began to purchase greater numbers of slaves from the Bight of Benin. These slaves accounted for about 30 percent of the Africans imported into Jamaica during the last two decades of the seventeenth century. The Popos were from this region, where they occupied the area extending from the mouth of the Volta River east to the Kingdom of Allada, straddling what is now the border between Togo and Benin. They were closely related to the Whydaws, who dominated the supply side of the late seventeenth-century European slave trade along the Bight of Benin. The short piece labeled "Papa" in Sloane's record was most likely meant to represent music from this region. Shifting political boundaries and competition from several other European nations made the supply of human chattels from this region, which was then becoming known as the "Slave Coast," too unreliable to meet Jamaica's increasing demand for long.[54]

Sloane preserved what appears to be first-generation music of transatlantic slaves from these three distinct African regions. By comparing the music with traditional African regional styles and placing it in the context of late seventeenth-century Jamaican slavery, a crucial missing link can be forged in the acculturation debates. Sloane recorded neither African American culture nor African culture in the Americas. Instead, his evidence tells of Koromantis, Papas, and Angolans using instrumental sounds and their voices to forge identities as Africans under the bonds of slavery in a harsh new world.

Early American accounts of African music are rare. Drouin De Bercy recorded one snippet played by soon-to-be-freed slaves in late-eighteenth-century Santo Domingo, but his transcription was not as meticulous as Baptiste's. Verbal descriptions are more common but less illuminating. In about 1640, Jean-Baptiste Du Tertre (probably not the same Baptiste as

Sloane's acquaintance) depicted a scene similar to that in Sloane for the French West Indies, except that real drums were used. A few years later, Richard Ligon described, in addition to drumming and dancing, the construction of an African-designed xylophone with wooden keys. In his late-eighteenth-century narrative of the Surinam slave revolts, John Stedman provided one of the best depictions of creolized African music in the American colonies, but even that is minimal and biased concerning the music itself. Stedman was more painstaking in sketching "their instruments of Sound," which were "not a Little in Genious," being "All made by themselves."[55]

In addition to Baptiste's transcriptions, Sloane wrote one of the most thorough descriptions of music and dance in the seventeenth-century Americas:

> [They] will at nights, or on Feast days Dance and Sing; their Songs are all bawdy and leading that way. They have several sorts of instruments in imitation of Lutes, made of small Gourds fitted with Necks, strung with Horsehairs or the peeled stalks of climbing Plants or Withs. Their Instruments are sometimes made of hollow'd Timber covered with Parchment or other Skins wetted, having a Bow for its Neck, the Strings tied longer or shorter, as they would alter their sounds. . . . They have likewise in their Dances Rattles ty'd to their Legs and Wrists, and in their Hands, with which they make a noise, keeping time with one who makes a noise answering it on the mouth of an empty Gourd or Jar with his Hand. . . . They formerly on their Festivals were allowed the use of Trumpets after their Fashion, and Drums made of a piece of a hollow Tree, covered on one end with any green Skin, and stretched with Thouls or Pins. But making use of these in their Wars at home in Africa, it was thought too much inciting them to Rebellion, and so they were prohibited by the Customs of the Island.[56]

How do Sloane's description and Baptiste's transcription relate to African music of the same period? This was an endpoint for region-specific African music and a point of departure for New World music. The Angola and Koromanti labels held different meanings for the musicians in Jamaica than they would have in Africa. Ethnic identity was still meaningful, but at the same time there was a good amount of syncretism among the musical cultures present. These two pieces of music are long enough to be analyzed in terms of language and musical style. Unfortunately, the piece labeled "Papa" is too short to appraise in the same depth as the others.

The relationship between the Sloane records and concurrent music in West and Central Africa can be explored in three ways. The music may be analyzed for similarities to the scales and rhythms of the traditional music of the area in Africa from which it is purported to have come. The instruments may be compared in a similar fashion.

The first piece of music, "Angola," shows four melodic traits more appropriate to the Akan region of modern Ghana (the Koromanti region) than to Angola. First, intervals of a third in the upper register of the piece occur three times as much as they would if intervals were distributed randomly. The pattern of thirds is common in the Akan region today, while neighboring areas use octaves or unison extensively. Second, intervals of a fourth, which would be more indicative of Angolan origin, are completely absent from the upper register. Third, the upper register uses a seven-tone scale common among the Akan. Finally, the piece is structured with a section of two-part polyphony, with its emphasis on the horizontal interplay between the two melodies rather than vertical "block-chording" prevalent in Western popular music. This horizontal presentation of harmony is another distinguishing characteristic of Akan music.[57] From the melody of the upper register, and the way the upper register relates to the lower register, it appears that it is of probable Akan origin.

In contrast, the bass register of "Angola" is marked by features more Angolan than Akan. The lower register uses a five-tone scale substantially different from its seven-tone counterpart in the upper register. Five-tone scales are rarely found in Akan music. The predominance of an interval of a fourth also indicates a culture other than Akan. However, pentatonic scales and intervals of a fourth are typical of—though not limited to—Central African music.[58] So while the upper register is probably Akan in origin, the lower register is most likely Angolan. Perhaps the European observers asked the Angolan rather than the Akan musician the origins of the piece.

The melodic form of "Koromanti" bears only limited comparison to traditional Ghanaian music. Akan music is noted for the predominance of intervals of a third, which, while not totally absent from this piece, are much less frequent than in "Angola," especially in the latter's upper register. "Koromanti" also makes extensive use of runs of consecutive seconds over a span of more than an octave. This has never been observed to happen among the Akan by the ethnomusicologist J. H. Kwabena Nketia, who notes that chains of seconds, whether ascending or descending, are usually "broken up after three, four or five or occasionally six steps by a pause or a change in direction."[59]

"Koromanti" is divided into three sections, each in a different tonality or "key." In each of the three sections, the would-be Koromanti musician uses a seven-note scale, indicative of Akan origins. The first two sections of "Koromanti" use seven notes, the third eight. The extra note in the third section was probably the result of an attempt by Baptiste to record microtones, which cannot be represented by standard European notation. Many African (as well as other) musical traditions make use of microtones in their tunings. These are notes that would fall between the keys of a piano.

An example familiar to Western audiences (albeit one of African ancestry) would be the bending of a string by a blues or rock guitar player to accent a note. Baptiste probably did not know how to deal with the microtones and rendered some of them as one note, and some as another. Correcting this would give a third seven-note scale, one that (probably not coincidentally) constitutes the typical blues scale in twentieth-century music.

The use of microtones is not common among the Akan, who show a preference for a seven-note scale based on the natural overtone series where the seventh interval is flatted slightly. This type of scale would not have caused Baptiste any confusion, as it resembles European "non-tempered" scales such as those produced by overblowing a flute, for example. The Angola region, which is known for its employment of microtones, is not known for its use of heptatonic scales. Although "Koromanti" contains several traditional Akan melodic structures, the musician used these features in unconventional ways. From the melodic evidence, it is likely that "Koromanti" is a creolized piece of music, fusing multiple styles into one coherent new one.

Sloane mentions the percussion: dancers with rattles set the beat over which a drummer improvised on a gourd or jar. Angolan and West African drummers both use this pattern, and it is still to be found in the Tambo and Koromanti drumming traditions in Jamaica, the former claiming Central African descent, the latter West African. The only distinguishing rhythmic evidence again points toward creolization. Nketia marks variety in the length of notes as a distinguishing characteristic of traditional Akan music, but "Koromanti" uses long runs of same-length notes, as does the bass register of "Angola."[60] This contrasts with the variety and relative non-repetitiveness of note lengths employed in the upper register of "Angola," which supports the notion of non-Angolan origins in that register.

Baptiste did not record the drum patterns, though he did capture the syncopation and polymeter in all three pieces—no mean feat for an amateur European musician untrained in African musical styles. It is a credit to his skill that he was able to record polymeter without altering it to force a fit with European expectations of meter. Sub-Saharan African drumming patterns have been extensively analyzed and classified by region.[61] Unfortunately, the actual drum patterns heard by Sloane and Baptiste were not recorded. They were probably the same or similar to patterns used when the musicians were in Africa.

Much African music is nondirectional in its use of time: melodic and rhythmic modules are begun and repeated; variation is supplied from combining, superimposing, and mutating basic modules into countless patterns. A piece may be very short or may be extended for hours. Musicians end a piece when they are finished rather than when it is finished. In European cultures, by contrast, time is traditionally interpreted to be moving in a linear fashion toward a goal: music typically has a climax and an end. Of course, there are overlaps between, and exceptions to, these two perspectives on time, but in general the African view is more cyclical, the European more linear. The "Angola" and "Papa" transcriptions probably each represent an "African"-type module that was repeated and elaborated. The subsections of "Koromanti" may indicate the same.[62]

Drums and Power

The sounds of *things* African were few in early America, yet African instrumental soundways, though transformed, survived to provide African Americans with one of the few free places they could craft in an often hostile and most unfree new world. A few West African musical instruments made the same harsh middle passage as the enslaved Africans onboard, probably getting better treatment than the human cargo. They were used for "dancing" the coffled Africans once or twice each day on the deck of the slave ship, a gruesome form of calisthenics to keep the slavers' shipment from atrophying. Once the slaver had disembarked the captive passengers, the instruments were perhaps kept aboard for another passage, maybe finding their way eventually into a cabinet of curiosities in old England, or just discarded. None would have been given to the human chattel about to be sold into a short brutal life on a Caribbean plantation or sent on yet again another passage of hundreds of miles to a North American plantation.

Although material objects did not make the passage, creolized West African instrumental soundways allowed the enslaved to craft autonomous agendas in colonial America, under even the most unfree conditions. Autonomous agendas are *not* autonomous lives. For Africans in the Americas, slavery precluded the latter in an obvious way. Yet they were able to craft sonic spaces that in some ways existed not in reference to slavery but to their own interests. Some initial work has been done marking out how West Africans and other marginalized people moved through complex communication networks that tied together early modern capitalist markets.[63] Instrumental music was a powerful tool in delimiting a covert set of knowledge, public only to those buried within that many-headed hydra. Planters, situated outside the inner workings of the economy they depended upon, knew that West Africans had ways of communicating through music, but they did not know how to stop them.

During 1930s, an ex-slave from St. Simon's Island, Ben Johnson, recalled from his childhood an old African man, Dembo, who was familiar with the traditional African uses of drums. Dembo used to beat a drum at funerals (and probably at feasts), but his master, the Yale alumnus James Hamilton Couper, banned the practice, ostensibly on religious grounds. Johnson said Couper did not want drums beaten around the dead. By the 1930s, the uses of drums for spiritual and festive occasions had seemingly ceased. But many of the coastal Georgian African Americans remembered the use of drums, or someone who knew how to make them, and a few admitted that the practice still existed. While drums may have been scarce, the practice of social representation through complex, culturally specific rhythmic patterns thrived in the work songs and hand-clapping patterns that accompany Sea Island spirituals to this day.[64]

African Americans had solid historical grounds to be reticent about their drums and drumming, reasons that stretched across centuries, continents, ethnicities, and racial divides. For example, the Capuchin missionary Girolamo Merolla described his reaction to drumming in 1682. He lived and worked in Songo, a Central African state about 150 miles southeast of Angola, part of a vast region of closely related societies. He wrote that drums were "commonly made use of at unlawful Feasts and Merry-makings, and [were] beaten upon with the Hands, which nevertheless makes a noise to be heard at a great distance." These drums, he continues, were also used for military signaling, for invoking the other world, and for sending off the dead properly. Merolla claimed that he often went to break up such "Hellish Practices, But the People always ran away as soon as I ever came up to them, so I could never lay hold on any to make an Example of them."[65] Although Merolla had definite ideas about controlling African practices that he found threatening ("Hellish"), he too had only limited agency. Whether in Central Africa or North America, Africans used their instrumental soundways to carve out locations of power, often in the face of concerted European resistance.

West and Central African horn and drum music often expressed state and institutional power. In a 1684 letter written from the Gold Coast town of Gross Friedrichsberg, Johan Nieman described the singing, drumming, and horn-playing that he heard there as "the most frightening and strange tones and dancing with the oddest movements in the world." Otto Friedrich von der Groeben, visiting the same area in 1694, heard horn music that sounded "like the shepherds in the villages in our country blowing the Christmas mass," although he does not make clear whether this assessment was meant to be pejorative or complimentary.[66] These seventeenth- and eighteenth-century horn and drum ensembles formed an elite within

Gold Coast societies, allowed by law (in Africa, not Jamaica) to play only for major political leaders. When captured in wartime, court musicians became royal slaves and as such could not be sold or traded by their new owners. To capture another state's court orchestra was considered a great accomplishment.[67] Probably very few court musicians were sold into Jamaican slavery; they would have been rescued or ransomed by their own people or kept as a prize by their captors.

From the beginning of the Atlantic slave trade, Europeans knew that drums were powerful tools of state among many West African peoples, but could not quite comprehend *how* this was so. English planters in the West Indies early associated African drum and horn music with mass uprisings of enslaved Africans seeking their freedom. Jamaican planters prohibited drums and horns as early as 1688. Barbados followed suit in 1699, banning drums, horns, or "any other loud instruments." Masters were to conduct weekly searches of slave quarters, and any of the named instruments found were to be burned under threat of a fine. In 1711, and again in 1722, St. Kitts passed laws which banned the slaves "from communicating at a distance by beating drums or blowing horns." In 1717, Jamaica codified its earlier policy forbidding "the gathering of slaves by the beating of drums and blowing of horns." A newspaper article from 1736 reported a foiled uprising in Antigua which involved "Coromantee" and colony-born factions of slaves. The Coromantee leader, Court, announced his intention to stage an uprising "in open Day-light, by a Military Dance and Show, of which the Whites and even the Slaves (who were not Coromantees nor let into the Secret) might be Spectators, and yet ignorant of the Meaning." The "meaning" was delivered by "Drums beating the *Ikem Beat.*" This plan was also found out, and many slave executions ensued.[68]

Planters passed laws against drums and drumming several times, and in various forms, indicating that their control was less than absolute. European fears were straightforward. They feared drums as loud signals that could lead men on a battlefield. Thus, they banned loud instruments, ignoring quieter ones in their laws. They understood only the soundways of military and state drumming that they shared with Africans: planters failed to comprehend how African Americans could represent themselves and their agendas in their music rather than just signal with it. These instrumental soundways, misapprehended by planters, proved to be powerful tools even without the instruments usually associated with them.

One particular type of West African court music—that radiating from a Kwa ethnic base centered in the region from Eastern Nigeria to Modern Ghana—was more than a set of signals. It functioned as an immanent and immediate means of representing and communicating ideas in a repeat-

able form, somewhat like a spoken language. Most Kwa languages are
tonal; that is, words can be differentiated on the basis of pitch change. Kwa
drummers, able to rely in part on pitch patterns, could produce musical
representations akin to a language rather than being a fixed corpus of signals. On the other hand, Mende and West Atlantic languages found to the
north of the Kwa regions, as well as Central African languages to the south,
are often tonal, but not "tonemic," that is, words cannot be distinguished
on the basis of pitch, but pitch still forms an aspect of correct pronunciation.[69] In these regions drumming still represented state power, but not
linguistically. Although most regions of Africa from which captives were
drawn had drumming traditions connected in some way to state displays
of power, its practice was strongest and most developed in the Kwa region
around modern Ghana—a situation of which planters, missionaries, and
traders seemed well apprised.[70]

Travel accounts from the Gold Coast also mention stringed instruments,
but these were folk rather than royal instruments. After a visit to the region
in 1602, Peter de Marees wrote of "small Lutes, made out of a block, with
a neck, like a harp with 6 strings made of rush, on which they [the Akan]

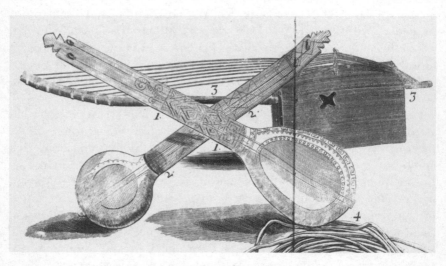

Fig. 2.10 Africans' musical instruments in seventeenth-century Jamaica. The lute in front (1)
is an Indian *tanpura* shown for comparison. The middle instrument (2) is a Central African
lute. The instrument behind the lutes (3) is an eight-string West African harp. The instruments were strung with wound fiber (4). Sticks (5) were used by enslaved Africans to clean
their teeth. Hans Sloane, *A Voyage to the Islands of Madera, Barbados, Nieves, St. Christopher, and
Jamaica*, vol. 2, plate 232. Courtesy of the Trustees of the Boston Public Library.

play with both hands," keeping "in good tune." Nieman reported that the Akan had "a sort of guitar, which they can play fairly well and sing pleasantly to."[71]

In Jamaica, slaves countered planters' suppression of African court music not only by direct resistance but also with the type of folk music recorded by Sloane. Drum patterns might be carried on in the rhythms of the stringed instruments or on smaller percussion instruments. Lutes and harps could be easily made; eyewitnesses remember these instruments being played as recently as the 1920s.[72] An eight-stringed harp was probably the instrument used to play the upper register in "Angola." Since this instrument could not be fingered, it could produce only eight pitches—the exact number required by "Angola's" upper register.

The lower register of "Angola" was probably played on one of the fretted lutes, which has noticeable Angolan attributes. The cross-hatching engraved on the neck of the middle instrument (the other lute is South Asian) is strong evidence of Angolan origin.[73] The order and range of the notes is such that they could have been easily fretted on this lute. "Angola" has another Akan quality. The vocal is in a declamatory call-and-response style, falling within the preferred Akan framework of a short vocal solo followed by a choral response and/or instrumental passage. This is an especially popular structure in modern traditional Akan lute music, although it is not particularly distinguishing, as many other types of African music also display an antiphonal pattern of some kind.[74]

The instrumentation in "Koromanti" is not as clear as in "Angola," but the layout of the notes suggests a keyed instrument such as a sansa or marimba rather than a fretted one like the two-stringed lutes. The use of an eight-stringed harp is precluded because such an instrument was capable of producing only eight pitches, while "Koromanti" requires sixteen. The marimba and the sansa were strange to European sensibilities. The Capuchin missionary Denis de Carli considered the marimba, or balofo, to be the "most ingenious and agreeable" of all Kongolese instruments. To build it, the Kongolese would make a wooden bow, binding to it "fifteen long, dry, empty Gourds, or *Calabashes* of several sizes, to sound several Notes, with a hole at top, and a lesser hole four fingers lower, and stop it up half way, covering also that at the top [of the gourds] with a little thin bit of Board." These thin boards were the keys the player would strike with "two Sticks, the ends whereof are cover'd with a bit of Rag." Playing the marimba produced a timbre that Carli thought "resembles the sound of an Organ, and makes a pretty agreeable Harmony." Ligon also remarked upon a mid-seventeenth century Barbadian slave's construction of a marimba, wrongly assuming it to be a new invention.[75] In Surinam, Sted-

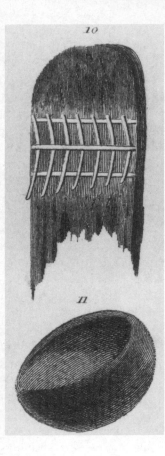

Fig. 2.11 A wooden *sansa* with an un-attached calabash resonator, as seen by John Gabriel Stedman in Surinam in the 1770s. The engraving by William Blake appears in Stedman, *Narrative of a Five Years' Expedition, against the Revolted Negroes of Surinam* (London, 1796), vol. 2, plate 69, detail. Courtesy of the Trustees of the Boston Public Library.

man characterized the sansa, to which he gave the Central African moniker *Loango bania,* as

> Exceedingly Curious being a Dry board on which are Laced, and kept Closs by a Transverse Bar, different Sized Elastick Splinders of the Palm Tree, in Such a manner that both ends are elevated by other Transverse Bars that are Fix'd under them and the Above Apparatis being placed on . . . a Large *Empty Gourd* to promote the sound/ the extremities of the Splinders are Snapt by the fingers, Something in the manner of a piano Forto & have the same Effect.[76]

Sansas, which are still found throughout the Caribbean today, usually have the notes laid out in alternate steps, one for the right hand and the next for the left, while the notes on marimbas most often run consecutively from low to high. These designs, particularly that of the former, facilitate the playing of long descending or ascending passages of seconds. Two-stringed lutes are ill-suited for this task, which was one of the main characteristics of "Koromanti." An additional reason to suppose a keyed instrument is that the lengths of the notes are fairly regular and short, indicating plucked or struck notes with little or no sustain. This is also in keeping with the nature of either the sansa or the marimba.

Although common in the Angola area, these keyed instruments were seldom used in the Akan region.[77] Sloane alludes to neither marimba nor sansa directly, although his mention of striking the mouths of gourds could be a reference to the resonators of a sansa, as he states that the gourds were struck with the hands, not sticks. Perhaps Sloane missed the sansa resting inside the more familiar-looking calabash resonator.

Ropes and Strands

When the instrumental and melodic evidence is woven together with regional African features as the organizing principle, two patterns of African ethnicity in the Americas emerge. In "Angola," two distinct ethnicities were combined but cultural boundaries were maintained in each. In contrast, "Koromanti" reflects an idealized conception of African ethnicity, one that has been remade in the Americas.

"Angola" and the corresponding illustrations imply a mixture of two discrete styles, with a Kwa—probably an Akan—musician playing harp and singing in the upper register while an Angolan was playing lute in the lower register. The intervals, scales, harmonies, instrumentation, and language of the upper register all point toward Akan origins and away from the Central African cultures indicated by the title. However, the lower register does match the Angolan identification of the title, as evidenced by its scales, intervals, and instrumentation. The two primary musicians—one playing an eight-stringed harp and singing the lead, the other playing the bass on a two-stringed lute—were positioning their identities in a way that allowed them to negotiate a whole that was new from parts that were not. This was the process of pidginization in action.

The most striking feature of "Koromanti" is the set of lyrics, which are identifiably Western Kwa, in semblance with the title. The words, however, cannot be translated fully or situated in a particular Kwa language with confidence. While languages may change somewhat in the course of three centuries, earlier and later versions are usually still mutually intelligible.[78] Once an analysis of the music is introduced, a Western Kwa identification becomes less certain. The ways in which the scales and intervals are com-

bined in a repetitive, consecutive manner, both melodically and rhythmi-
cally, not only point away from an Akan origin; they imply the use of instruments that were not common to the region either.

Mimetic explanations of African ethnicity break down in "Koromanti."
Traits from unrelated cultures cross categorical boundaries, blending in
ways that the music in "Angola" does not. Ethnic identity in the Americas
had become idealized in "Koromanti," but not in "Angola."[79] The external
cross-cultural mixing taking place in "Angola" had become internalized in
"Koromanti." The result was a creolized ethnic identity. "Koromanti" was
probably the expression of a seasoned slave whose African ethnic identity
had been formed wholly or mostly in the Americas. To such a person, Koromanti ethnic identification would no longer signify a subregion of Western Kwa culture but a loosely bundled set of associations centered on West
Africa that helped one operate in the context of a new world.

The instrumental sounds in Sloane's musical description address issues
raised in the acculturation debates. Historians have most often drawn their
conclusions from the more numerous records of fully developed slave populations or, in many cases, from post-emancipation documents. The importance of Sloane's account is its early date and specificity. A glance at the
titles alone of each piece of music calls into question any contention that
all was lost in the process of enslavement. Yet Angolans and Koromantis
did not simply retain their African cultures intact; a process of interchange
and experimentation was definitely taking place in the music. Mintz and
Price's hypothesis—that purposefully randomized slave crowds resulted in
a cultural leveling process—is not borne out either. The ways in which
these negotiations were carried out were still ethnically identifiable in
1688.

If African cultures were not destroyed, replicated, or leveled by transit
to the Americas, then how can description of what happened, as exemplified by the music, be approached? Pidginization provides a key. When
adults acquire a second language, they never learn it as well as a native
speaker. When adults acquire a second language, they will always be recognizable as non-native speakers, given away by some slight difference in
pronunciation or stress. Pidginization is the same process, but the language must be reinvented from what is available; there are no native speakers. As a result, none can ever be quite at ease with the tentative music,
language, or culture of the first generation. This is illustrated by "Angola's"
uneasy blending of two different musical styles, each part retaining its individual distinctiveness. Both musicians were using their own particular
cultural knowledge to produce a suitable musical expression. In all likeli-
hood, neither the musicians nor their companions found the results fully
satisfactory.

The first, or pidgin, generation had to negotiate and compromise ad hoc to reach any shared cultural understanding or sense of community. Later, creolized generations acquired this makeshift culture as native, expanding and formalizing it—making it fully their own, as exemplified by "Koromanti." They molded the creole cultural language that later immigrants would have to learn as a cultural second language. The resulting creole form elucidates the lack of explanatory power of most retentions: they are not sacrosanct traditions handed down reverently through the ages but provisional solutions that only became native when a new generation acquired them as part of a cultural first language.

The African diaspora to the Americas may be thought of as the unraveling of a number of individual twines from different ropes and their recombination into new ropes. Thus, pidginization (both broadly and narrowly defined) occurred. As time passed, new groups of twines, some with new kinds of strands in them, coalesced out of the relatively chaotic pidgin stage to weave new ropes with new and discrete cultural definitions. This was the creole stage. Today, African Jamaican music and language exist in a mature creole culture. Only its history distinguishes it from a noncreole culture. Most historical research into African American cultures has focused on creole generations. Descriptions such as those in Hans Sloane's account give historians a glimpse into the earliest workings of African cultural transmission (to repeat Bilby's words) "in action at the individual level, where conscious creative decisions (as well as unconscious adjustments) were made." This was a world ordered by decisions and processes much more individualistic and unpredictable than those of later, more stable communities, but the latter owe their existence and contents largely to the results of the former.

South Carolina

During the eighteenth century, evidence of African American instrumental soundways in North America begins to emerge. Although North American rice planters knew well the attitudes of Caribbean planters toward African drums and drumming, the instruments were not banned at first in South Carolina. In fact, African slaves were often used as drummers in the militia. Peter Wood suggests that African militia drummers were so prevalent in South Carolina that the job was seen as unattractive by "race conscious Europeans."[80]

Rather than banning drums, South Carolina and Georgia rice planters simply did not purchase many Kwa males who—being preferred in the wealthier and more established sugar colonies—were scarcely available in the low country anyway. The early accommodation to Africans' drumming in South Carolina was uneasy, though. The planters feared the power of West and Central African drums and drumming. In 1730, according to a Charleston planter, a group of slaves "conspired to Rise and destroy us" at a dance that featured drumming. The alleged revolt was discovered and quelled before the slaves were able to issue a call to arms, however. This drumming was associated with members of particular Kwa nations, who in turn would have translated the call to their allies, probably via a pidgin or creole interlanguage.[81]

Dances with drums were not the only way that enslaved Africans linked instrumental soundways to martial skills. A notice in a *South Carolina Gazette* from 1733 offered a £10 reward for the return of Thomas Butler, who had run away from the Vander Dussen plantation upriver from Charleston. Butler was known in the area as "the *famous* Pushing and Dancing Master." For an owner, especially one as intolerant as Vander Dussen seems to have been, to refer to a slave as "master" seems ironic and unusual.[82] It is doubtful that Butler was the master of any dancing skill of which his owner also partook. Nor is it probable that planters sent their children or slaves to such a master to learn dancing—especially not a style which included "Pushing" as a major feature. This was not ballroom dancing. Butler was, however, not only a master of "Pushing and Dancing," he was *"famous"* for it.

Butler's skill was most likely not of his own invention. John Storm Roberts discusses the mid-nineteenth-century popularity of a Brazilian form of musical martial art called *capoeira de Angola,* which was practiced by young men, often from Central Africa. The art could best be described in two words as pushing and dancing. *Capoeira* uses musical bows, which are percussive string instruments, to set a polymetric tempo that disciplines the movements of two dancers who combat each other in a highly ritualized and graceful manner. The musical bow is a much quieter instrument than a drum and can be quickly made from a flexible green tree limb, a length of string or cord, a small stone, a gourd, and a striking stick.[83] The sound is more percussive than melodic. Perhaps the roots of this pushing and dancing martial art lay in trying to keep traditional combat skills honed with no weapons available and drums proscribed. Such intensely purposeful dancing, performed by a master, would surely bring about the attention of planters, but not necessarily their comprehension of what the music and dance were representing.

Angolan and Kongolese warriors in Africa also had a form of "Pushing and Dancing." Hand-to-hand combat was still a viable military skill in the early eighteenth century, though finally being superannuated by firearms—for which, noted the feared warrior Queen Njinga, "there was no remedy." Kongolese and Angolan techniques of unarmed combat were learned in the form of a martial art set in time to drum music. In short, the skills were encoded in a form of dance. Not all soldiers learned these techniques. Specialists, called *imbare* (singular *kimbare* or *quimbare*) and often drawn from slave populations, were recruited to learn the art. According to John Thornton, this specialized form of dance, called *sanga* in the Kikongo language, and *sanguar* in Ndongo, valued hand-to-hand combat skills, the use of sticks and other weapons, as well as "the ability to twist, leap, and dodge to avoid arrows or the blows of opponents." The skills, which brought renown to the *imbare,* were displayed at public exhibitions that impressed not only Africans, but Portuguese, Italian, and Dutch observers as well. Thornton notes how a Kongolese state delegation in Brazil amazed observers there with an exhibition of leaping and fighting skills in 1642.[84]

By the eighteenth century, Central African armies had developed mass-mobilization infantry tactics as a result of a century of civil war. The importance of *sanga* as a military form was waning. The Americas, however, were rife with evidence of its retention. It may often have been more important as a ritual than as a military tactic, but it may have yet had its uses in the latter arena. Ritualized stick fighting and dancing like that found in Central Africa persisted all over the "new world." One of these dances, called *kalinda,* was a highlight of Caribbean slave festivals, although viewed ambivalently by planters. Brazilian slaves kept their unarmed combat skills honed in *capoeira.* A related martial art/dance form, *maculelê,* existed alongside *capoeira* in Bahia. In it, two dancers used sticks called *grimas* as musical instruments and as weapons against each other, both at the same time: to miss a beat was to receive a blow. In Cuba, the Kongolese tradition of music and dance was also closely associated with military traditions. In the United States there is a tradition of "knocking and kicking." Even baton twirling takes on wider significance, for as a relative of *kalinda* and *rara,* it can be situated as a part of this complex system of musical, social, religious, and military instrumental soundways.[85]

In a world where overt possession of weapons was limited or banned, the representation of social knowledge and military skills via rhythmic patterns of drumming and movement could be highly valuable. It was no coincidence that two of the most-frowned upon activities in which a coastal lowlands slave could engage after 1740 were reading and particular forms of music—namely, the loudest forms, drums and horns. But in trying to control the knowledge that Africans had access to, Europeans only considered their own soundways. In their worries about loud signaling instruments, the planters missed the purpose of Thomas Butler's quieter art—and many other African military soundways. They even graced Butler with the moniker "famous master."

Three years after the Antigua conspiracy, and six years after Thomas Butler's escape, South Carolina planters' worst nightmare came true, "an intestine Enemy the most dreadful of Enemies." About twenty slaves, all sharing a common Central African cultural background, "surpriz'd a Warehouse belonging to Mr. Hutchenson, at a Place called [Stono]; they there killed Mr. Robert Bathurst and Mr. Gibbs, plunder'd the House, and took a pretty many small Arms and Powder."[86] The slaves—unarmed—had effectively taken over a small arsenal. Guns and ammunition were no doubt their immediate aim, but how did they obtain them? Perhaps the last thing that the two armed sentries experienced was the combat version of Thomas Butler's "famous Pushing and Dancing," deadly hand-to-hand fighting tactics that the slaves were able to maintain as a form of musical dance even under the direct observation of planters who feared just such an enemy.

After thus arming themselves, the slaves marched southward with "Colours displayed and two Drums beating." When their ranks had swollen to between sixty and one hundred slave defectors, they stopped, still not far from Charleston, and "set to Dancing, Singing and beating Drums" for the purpose of calling more slaves to join them. This drumming, true to Central African traditions, was in the form of an announcement, a signal, rather than a Kwa language. By this time, the planters had recovered sufficiently to respond with force. A pitched battle ensued in which more than twenty whites and twenty slaves were killed before the slaves scattered. Many runaways were captured and shot during the following weeks, but the insurrection was not considered quelled for at least a month.[87]

The evidence at Stono points to ways of doing battle that were not pan-African, much less universal. The manner of fighting and the indications of how power was expressed through music point to Central Africa, but the expressions themselves were tailored to the conditions of slavery. The escaped slaves suited Central African soundways to a particularly American situation. Attention to these instrumental soundways helps us discern how Africans from diverse regional backgrounds came to understand each other and act in concert.

Local planters had been quibbling relentlessly over the codification of new slave restrictions until the results of the Stono insurrection made cooperation imperative. In 1740, the new slave code was rushed through the Assembly. At the same time, South Carolinians stepped up hostilities toward the Spanish in St. Augustine, who offered freedom to any slaves who could escape there. Among its strictures, the new slave code prohibited "wooden swords and other dangerous weapons, or using or keeping of drums, horns, or other loud instruments which may call together, or give sign or notice to one another of their wicked designs or purposes." Dena Epstein has observed without elaboration that musical instruments were classed in the same category as dangerous weapons.[88] Perhaps the wooden swords were the sticks used in *kalinda* and *maculelê*. The planters' reaction, with its focus on specific material objects, underscored their lack of understanding. The new restrictions were analogous to taking pens, paper, and books from the literate. Such an action could have definite effects, but it would not render the literate population illiterate. The planters' worst (and perhaps only) fear was that of a power they clearly apprehended, but had no way of comprehending.

African American drumming at Stono was an act of self-determination based on an autonomous agenda. It consisted of creolized West and Central African instrumental soundways of transforming a deep—perhaps universal—human belief in the pursuit of freedom (in the sense of autonomy for one's self or group) into a unique expression suited to the exigencies of American slavery. To speak of only "slave" resistance and accommodation in this case seems an odd twisting of the facts. In the case of drumming and music, the planters were the ones reacting defensively to ways not their own. Perhaps that is why Epstein claims, arguing from a different perspective, that "[i]rrational though it may have been, the fear of drumming as a signal of insurrection persisted up to the outbreak of the Civil War."[89] Indeed, South Carolina's laws against drums stayed on the books until Emancipation, and enforcement seems to have been fairly thorough. Couper's nineteenth-century ban on beating drums around the dead thus rested on something much broader than a simple religious belief. It rested on a long tradition of European resistance and accommodation to African ways. Though slavery was harsh, the planters' power was by no means absolute. The Hegelian master/slave dialectic begins to look like very thin description when there was so much more to African identity in North America than could be encompassed by the term "slave."

Fiddles and the *Jali* Tradition

After 1740, mentions of slaves playing drums virtually disappear from colonial records in South Carolina and Georgia. Curiously, drums seem to have been replaced by fiddles. From a single runaway violinist before the Stono uprising, the number of escaped lowlands fiddlers reported in the *South Carolina Gazette* steadily increased during the years before the Amer-

ican Revolution and then abruptly disappeared during the war, with the
next runaway fiddler not being noted until 1790.

Why was there an upsurge in runaway fiddlers? Violin playing provided
slaves with access to some key aspects of planter culture. Charleston was
noted as a musical center before the Revolution. Violins were instruments
of high culture to the planters, and possession of a musical ensemble was
a sign of status. Violinists were in demand for dances and entertainment.
Well-known in European folk and elite traditions, the instrument was not
thought of as a threat like drums were. In addition, slaves with experience
on the instrument could be hired out, bringing extra income to their
owners, and occasionally to themselves.[90] Such slaves would have access to
casual conversations of the planters, no doubt a source of valuable information. More importantly, they would have an amount of local mobility.
Together, these two job features provided key opportunities for potential
runaways, opportunities not available to field hands.

Fiddling must be learned; the violin requires guidance, practice, and
skill even to be played in tune, much less played well. How did slaves come
to possess proficiency on the instrument? The skill had to be learned at
some point. It is doubtful that many plantation owners would afford the
double luxury of a paid white violin tutor for a slave while at the same time
losing valuable labor time. Violin lesson books did not appear in the
colonies before the 1760s and were probably not widely distributed until
well into the nineteenth century. Such manuals would have been incom-
prehensible without both musical and verbal literacy. Literate slaves would
be more likely to possess religious texts or a hymn book with no musical
notation provided by an itinerant evangelist than an expensive and scarce
violin manual.[91]

Slaves most often learned the art of fiddling from each other. An eighteenth-century description from Santo Domingo maintained that "many
[slaves] are good violinists. That is the instrument they prefer. Many certainly play it only by rote, that is, they learn by themselves, imitating the
sounds of a tune, or they are taught by another Negro, who explains only
the position of the strings and the fingers, with no thought of notes."[92]

Novice fiddlers anywhere had to learn from someone who already knew
how to play tunefully. If enslaved violinists learned from each other, then
some of their knowledge must have come from Africa. The only part of the
slaving areas where bowed instruments were prevalent was the Mende/
Western Atlantic region, the same area where the rice cultivators preferred
by the coastal planters lived. This preference became even more pronounced after the Stono uprising, as Central Africans were no longer de-
sired.[93]

Information from runaway notices on nine slaves from the rice planting region between 1730 and 1790 shows most of them displayed some potential status marker. None, however, could be ascertained as African-born. Where were the Mende/West Atlantic violinists? Their playing and teaching took place mostly in a world of which Europeans only skimmed the surface. Although low country slave owners may have preferred Mende/West Atlantic African-born slaves over other African ethnicities for noncultivation jobs, American-born slaves were more generally preferred for these jobs and any others that required substantial contact with planter society. Mende violinists would not have had the same access to the valuable knowledge that noncultivation jobs provided. Furthermore, planters were less likely to know about the particular musical skills of their fieldhands, so even if Mende violinists ran away, their musical abilities might not turn up in the advertisements for them.

Planters found American-born, or "creolized" slaves, to be more predictable than African-born slaves. They were also less prone to running away—unless perhaps they had acquired some useful knowledge about how to get by, as it seems the runaway fiddlers had. Having grown up on the plantations, creolized slaves were much more acculturated into that way of life than immigrant Africans could ever be. They learned the music that Europeans wished them to play with the same facility with which they learned the methods and techniques of African teachers. They undoubtedly knew more about planter ways than first-generation African immigrants, both from the greater propinquity that their creole status was likely to provide and from the plantation being the site of their natal culture.

Expressing Anglo-American music in creolized African ways produced something that could accommodate slave and planter communities alike. One example of this would be the African-styled "jigs" and "reels" that began to be found during the eighteenth century. The best record, albeit a later one, was a recollection of youth written in 1876 by Henry W. Ravenal, in which he was invoking an older world, one that was fading even when he was a child. He wrote of Christmas festivities of his boyhood at his family's South Carolina plantation home, which the family had built in 1716:

> The jig was an African dance, and a famous one in old times, before more refined notions began to prevail. However, it was always called for by some of the older ones who remembered its steps. . . . For the jig the music would be changed. The fiddle would assume a low monotonous tone, the whole tune running on three or four notes only (when it could be heard). The stick-knocker changed his time, and beat a softer and slower measure. Indeed, only a few could give the "knock" for proper effect.[94]

This was not only African-derived music, it was probably a re-invention of the drum music so feared by the planters: though the thing (drums) and the form (drumming) were banned, an underlying value of the enslaved musicians (public representation) was expressed in a new way that was simultaneously (but not synonymously) European and African. These jigs would have been strange and unfamiliar to a Mende fiddler, as much for the Central African–derived dancing as for the soundways adapted from the planters. African American instrumental soundways helped create new, American components to the identities of the planters and the enslaved alike.

The low repetitive monotone of the violin, using only three or four notes of indefinite pitch, could have easily been an encoding of a banned drum style. Throughout West and Central Africa, drummers play with two hands, two sticks, or most often, one hand and one stick. In the last configuration, the drummer plays with one hand open and a drumstick in the other hand. By manipulating the tension of the drumhead with the open hand, a drummer could produce a number of distinct pitches that would be repeated and built up into a rhythmic pattern. Similarly, the fiddler's "three or four note" repetitive figures were played with an open left hand controlling the pitch and a stick, the bow, in the right. Bows, when bounced on a string, even respond (tactilely but not audibly) in much the same manner as sticks against drumheads. Evidence from Jamaica shows that "Kalinda" musicians even held their fiddles as if they were drums. Central Africans have many stringed instruments that are struck by sticks to produce changeable but indefinite-pitched percussive sound. Among them are the musical bows used to accompany *capoeira*, and another instrument shaped much like a violin with no strings, both of which can be traced to Central African sources.[95]

Sometimes while a fiddler played, a second person would take a sturdy pair of straws—or in one case, knitting needles—each about eighteen inches in length, and, facing the fiddler's left shoulder, strike the strings of the violin between the fiddler's bow and his left hand. The practice of having a second musician play percussion on a stringed instrument was prevalent in Latin America, too, where the percussionist played on the wooden parts of guitars or "creole harps" with sticks as another player used the strings. This treatment of a single instrument as two functionally discrete instruments also has Central African precedents. The "beating straws" technique has even found its way into white string band music.[96]

Again, it is the soundways, not the instruments (nor, in the case of white string bands, even the people playing) that illustrate what is African in creole cultural situations. Both Central African and Mende ways could be expressed simultaneously—in a way that Europeans understood differently as their own. The underlying value placed on music might not be a very illuminating Africanism. The expression of that music on violins played with bows, hands, knitting needles and straws was definitely original, an act of creation designed to meet the challenges of particular situations. But the soundways—the transformations of widely held beliefs about music into innovative instrumentations—can be traced to Central African, Mende, Kwa, and European ways of playing, discernable even when so thoroughly intertwined as they were in eighteenth-century South Carolina.

While drums were banned, the violin functioned well for quietly representing African drumming traditions that were so feared, but little understood, by planters. The polymeter rhythms of banned drums were stored in the distinctive pulse of the stick knockers and the fiddler's three- or four-note rhythmic pattern. In order not to give away their purpose, the patterns were beaten softly, as the planters feared only the loudness of instruments, showing no comprehension of the music's other ways of representing power.

The jig itself, when danced by whites, was always done in pairs, and took place in the center of a ring of people. A woman would enter the ring, doing a shuffling dance while gracefully waving a handkerchief over her head, and a man, again in Ravenal's words, would follow with "his whole soul and body thrown into the dance. The feet moved about in the most grotesque manner. . . . It was hard work, and at intervals of five or ten minutes, he was relieved by another jumping into the ring with a shout and shuffling him out."[97] The mock-confrontational shout of the entering male, and his "shuffling" of the other man out of the ring again evokes the ritualized combat of *capoeira* and the image of Thomas Butler's pushing and dancing. Perhaps the renowned slave would have enjoyed the irony of these pushing and dancing masters, unknowingly imitating the most deadly aspect of the music they sought to ban.

Violins obviously did not simply substitute for drums in the years following the Stono Rebellion. Creolized fiddling had a different set of capabilities for representation than did drums. Stringed instruments, whether bowed or plucked, were part of the *jali* (griot) tradition, which existed throughout the slaving regions of Africa, but generated from a hearth area in the Mende/West Atlantic regions—as opposed to the court drumming traditions that were mostly developed further south, in the Kwa and Kongo/Angolan regions. Thus, in the playing of jigs and reels on a classical European instrument, the violin, we can listen in on soundways that have roots simultaneously leading back to northern, western, and central sub-Saharan Africa as well as Europe.

The court tradition, which manifested itself in the drumming and dancing that so intimidated planters, was a means of directly representing and
displaying power to a public that understood its meaning. Drummers and
dancers were agents, representing an immanent fighting or political force.
Only insiders knew the meaning of the drumming message, though outsiders apprehended its power. Court drumming was an ephemeral, in-
stantaneous means of mass communication and representation, perhaps
the original form of broadcasting. Like a voice, as soon as the representation is uttered it is gone. But also like a voice, its expressive power extends
beyond the semantic content of the words alone. Like a broadcast, the volume of the drumming disembodied the sound, reaching beyond the face-
to-face.

Whereas court music represented power, *jali* songs described and explained it. The music of the *jali* tradition, to continue the analogy with
other media, is more like a text. It is an editable, manipulatable, analyzable
medium that can be recalled in the same form. The songs were not documents of the past so much as a means of encoding information. Usually
what they encoded was some sort of legitimation of or recipe for power and
its use. The way violins were used was as a *jali* form of storing powerful traditions, namely, court drumming patterns and the rhythms of military
dance. These stored forms could be reconstituted as direct manifestations
of power. They were an effective way of transmitting knowledge across
time.

"Turkish" Military Music in the Revolution

Descriptions of music during the American Revolution show that slaves
from the coastal lowlands were able to maintain their aptitude in drumming throughout the three and a half decades since it had been prohibited. The number of runaway African drummers from South Carolina that
were noted in print boomed from one during the previous forty years to
twenty between 1775 and 1780. All were Charleston slaves; all but one, the
"Negro Bob" who drummed for the South Carolina revolutionaries, joined
Hessian regiments that promised freedom in exchange for military service. At least eighty-two people from the colonies joined the Hessian forces
during the revolution. Of these eighty-two, fifty-two were drummers, and
thirty-five of the latter were black. Twenty-seven of the recruits, or about
one-third of the total, were from the Charleston plantation area. Twenty-
four of the Charleston recruits were black, of whom nineteen were employed as drummers, two as fifers and three as laborers. Only one of the
Hessians could be identified as African-born.[98]

Military drummers for the American revolutionaries were drawn from the rank and file. Their drumming was used mostly for sending field directions, and was in a state of disarray for much of the war. No particular facility on the drums was required to become a military drummer for the revolutionary forces. The main task was to send loud, simple coded instructions by means of rudimentary drum patterns. If Hessian troops had operated under the same standards as the revolutionaries, the reason for the large number of African drummers joining them could simply be written off as an interest in doing something that had been previously prohibited.[99]

Europeans, however, considered German military bands to be the best in the world from the 1750s until the turn of the century. German military units were participants in the craze for "Janissary" music that had been slowly sweeping westward through Europe from 1720 onward. However, the British did not adopt the style until the 1790s. During the American Revolution, English regiments continued with traditional fife-and-drum field units and *hautboy* military bands as the norm. The skill requirements for drumming in such a corps were not much different than those for American drummers.[100] This explains why men skilled in African drumming traditions found such ready positions in Hessian rather than in English or American regiments.

In theory, the "Janissary" style favored by the Hessians was derived from Turkish military music. For instruments it used several large drums, tambourines, and high-pitched flutes and reeds. But Europeans confronted Janissary music not in alliance, but as enemies, so the borrowing was often second-hand. In Germany, Africans became the preferred musicians for Janissary corps, especially as drummers. They were acclaimed as such and changed the drumming ways from the Turkish to what a regimental leader labeled "modern cross-handed drumming." They were dressed as flamboyantly as possible, and their marching was actually a cadenced dance that drew on the same sources as *kalinda, rara,* and baton twirling. It involved leaping and contortions as well as the throwing and catching of drumsticks and the adroit handling of batons and jangled sticks, all skills maintained by the culturally conditioned transformations of Africans' military values. Regiments would compete to have the best and wildest Janissary units.[101] Virtuoso skill was a requirement and could not be had on short notice. It took years of practice. It was exactly these roles that creole Africans from the low country stepped into when they joined Hessian forces.

The parallels between the "Janissary" performance and the violin music described above include stick work, agile dancing, rhythmic virtuosity, and strict adherence to time. The similarities to African instrumental soundways in court and military music include all of the above plus the court function of immediate representation. Much like Kwa court music, Janissary regimental performances communicated an immanent force. This function was in marked contrast to the *Jali*-like performances of African American fiddlers. But without the encoded "text" or "recipes" that were stored and represented in jigs and fiddling, creolized enslaved Africans would have been less likely to fill spots as Hessian drummers when the opportunities arose to "read these texts aloud" as displays of a present power.

There was more to African culture in the low country and elsewhere in the Americas than that in the institutional confines of slavery. Attention to soundways gives us access to the covert publics that Africans in the Americas used to carve out some sort of autonomous space for themselves. It was a world not only in resistance to or accommodation of the slaveholders, even though from the planters' perspective that might be all that could be seen. A close study of soundways of which the planters themselves were often unconscious reveals that the world they thought they owned was not always what they thought it was.

African culture in the Americas was more than simply a reaction to bondage, but it was not a simple transfer of the "African" to "America" either. Mende fiddling and *Jali* ways, Kwa-based court drumming and tone languages, Central African martial arts and percussion instruments, European-derived music and dance: all could be woven together seamlessly within a single expression uniquely suited to an American context, whether in pushing and dancing, jigs and reels, "Janissary" music, or—in the nineteenth and twentieth centuries—Sea Island spirituals. Focusing on the ways people transformed their values and beliefs into sonic expressions allows us to listen in on this creolizing world, even through the muffled records left by planters. The stories of how these transformations themselves changed over time provide us with keys not only to the history of African American identities, but to all the many American identities that have long struggled both to comprise and pull apart the very notion of "America."

In this chapter we have explored how instrumental sounds have been used to construct group and individual identities among European Americans, Native Americans, and African Americans. Only active sources of sound have been considered. Another type of instrumental sound shapes sounds rather than produces them. It is to these acoustical spaces that we now turn.

CHAPTER THREE

No Corner for the Devil to Hide

As both a musician and someone who studies sound, I have the odd habit of clapping once sharply or speaking in a loud staccato voice or humming to no one in particular when I enter what appears to be an acoustically interesting space. I was glad, then, that the Birmingham meeting of the Society of Friends was just about to start their worship when I asked to go into the old hexagonal-shaped schoolroom: I was given permission to go alone. The room responded to my claps and vocal probes with a "live" sound because the ceilings are made of hard plaster and shaped in such a way that sounds bounce immediately back with little in the way of complex echoes typically produced by the high, steep, two-sided ceiling in a gothic chapel. Today, shape-note singers, with their emphasis on the participation of all and their accentuation of the mid-range timbres of the voice, value these old Quaker halls for their participatory acoustics. Later users of the room, annoyed by the same acoustics, installed hooks in the ceiling from which to hang thick, sound-absorbing drapes to muffle the reverberation. The "live" sound of the undraped room was akin to the slap-back echo popularized on old rockabilly and rhythm and blues records from the late 1950s, but when I was there it sounded the same as it and other hexagonal Quaker schoolhouses and meetinghouses had sounded for centuries. The acoustics amplified everyone's voice, with no echoes building up anywhere because the shallow-ceilinged, obtuse-angled rooms had, according to one folk explanation, "no corner for the devil to hide."

Τhis chapter considers what eighteenth-century encyclopedists called "catacoustics," the study of how sound was instrumentally projected, reflected, dissipated, and otherwise manipulated once it had been produced.[1] By manipulating reflected sounds, early Americans added layers of meaning that enriched and reinforced deeply held beliefs. They carefully attended to the audible world when they created and shaped public—and in some cases, not so public—spaces. Acoustical spaces reflected the beliefs underlying social order as well as vocal and instrumental sounds. Thinking about acoustics, we can still hear the echoes of those social orders and begin to notice how people created and maintained ranked points of contact within their communities and nations, across divides of ethnicity, race, gender, and class, and (perhaps most important) between visible and invisible worlds.

Acoustic spaces and the quality of the sounds made therein are remarkably durable even though the particular sounds are ephemeral. Anyone who has ever tested the reverberation in an old high-ceilinged church knows the power and durability of acoustic design and the shape of the sounds made within it. Our everyday acoustics are now filtered through electronic amplifiers and speakers, giving us soft voices at loud volumes and problems of distortion, feedback, and tone unfathomable in the seventeenth century. And in many ways, it is difficult for people today to imagine a world where sounds with no visible source were necessarily other-worldly: our radio, television, and film voice-overs, as well as music recordings, routinely fragment sounds to the point that disembodiment is mundane. Yet the gap between our worlds and theirs can—at least in part—be breached, as a study of church acoustics makes clear.

European Church Acoustics

In order to understand Anglo-American acoustical soundways, we must begin with a baseline of European church acoustics. While the two traditions are parallel in many respects, we need to pay attention to the fact that colonists did not have to face the preexisting institutions and acoustic spaces with which Europeans contended during the Reformation.

Older church designs emphasized high ceilings made of hard, sound-reflecting materials. The priest stood within a semi-enclosed wing of the church, the chancel, facing the altar, his back to the chancel screen separating the altar from the rest of the churchgoers in the nave. The priest was nearly invisible to his auditors. He spoke and chanted in Latin, indecipherable to most from the outset. Language, however, formed only the last barrier to comprehension. Facing away from the congregation, the priest's voice never carried directly to listeners in the nave. It began its trip toward them as an echo, reflected several more times before reaching any ears.

While the medieval chancel is often negatively construed as an impediment to vision and acoustic clarity separating priest from congregants, it can also be considered a beautifully executed, very large musical instrument, somewhat like the body of a lute. The priest's voice provided the initial signal, like the plucked string of the lute. Unlike a lute string, however, the priest's voice was located inside the body of the instrument rather than

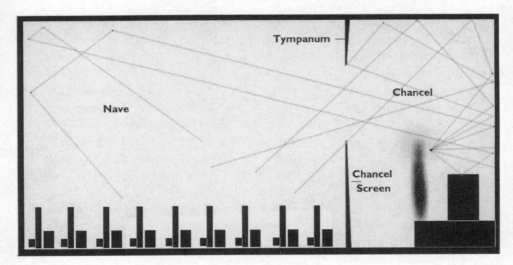

Fig. 3.1 Acoustics of a medieval chancel. Drawing by author.

outside, and it was carefully directed toward the back of the instrument, the concave eastern wall of the chancel, rather than being a diffuse signal such as that from a lute string. The chancel walls collected and directed the sound forward to the listeners like the back of the lute's body. En route, the signal encountered a set of vertical barriers. On the floor was the chancel-screen, topped with a crucifix or rood. The chancel screen was often carved so that people in the nave could partly see through it. While it made the priest's actions vaguely visible, the chancel screen's open tracery or perforations acted as an acoustic baffle, muffling and deflecting the floor-level sound waves emanating from the chancel. This would make the sound that did escape seem to come more from above than across. Directly above the chancel screen was an opening occupied by the rood, through which sounds passed more freely, something like the sound holes of the lute. Above that, hanging down from the ceiling, was the tympanum, named after a drum head or the eardrum. In effect, the tympanum acted like the sounding board of our hypothetical lute, vibrating when struck by the signal but also bouncing sounds back into the chancel until they were directed out of the opening where the rood was situated. Taken together, the chancel and its parts constituted a sort of reverberant sound amplifier. Because the signal was already reverberating before it left the chancel and because a chancel is much larger than a guitar, the sound emitting from it at any moment was a compendium of echoes, the sources of which overlapped in time much more so than those coming from a lute.[2]

Once this complex signal reached the nave, where the congregants were seated, the crosslike construction of medieval churches bounced it around more, creating cascades of echoes. The high, acoustically reflective ceilings added sonic power to the priest's voice, reverberating and reinforcing it, while at the same time further muddying it with the echoes that constituted the nave's reverberation. The steep incline of the high ceilings increased the reverberation time. Long after the first echoes faded, the last ones would still be escaping from the cavernous ceilings of a typical church. Sounds bounced around echo upon echo upon echo rather than reaching the listener's ears all at once. This created a powerfully moving effect, one that amplified the voice and enriched the tone, but at the cost of clarity. Acoustician Hope Bagenal somewhat derisively describes the emphasis on reverberant sound found in Catholic churches as the "acoustics of the cave," comparing it with the acoustics of the open air exemplified by Greek amphitheaters and, implicitly, Protestantism.[3]

The Reformation changed the acoustics of existing churches. Graven sounds as well as graven images had to be removed. Clarity of voice rather than fullness became the goal. The *Book of Common Prayer* instructed Anglican ministers to speak from the place where they would be heard most clearly. The foremost late-seventeenth-century British church architect, James Wren, took as his guiding design principle the creation of buildings where the minister could be seen and heard clearly by all. Martin Bucer had mapped out this position a century before, and it was adopted in the second edition of the *Book of Common Prayer* in 1552, so Wren's conservative position was hardly an innovation.[4] The pulpit and reading desk directed the sanctioned voices of authority emanating from the reader and the minister directly across the room to the ears of the congregation, amplifying the signal in the process. The first and loudest sound was the original, unreflected signal. The reflecting surfaces of the pulpit were so close to the speaker that the reflected sounds were perceived as part of the original signal, slightly "fattening" it and giving the impression of amplification rather than gothic reverberation.[5]

The flow of this direct signal was nonreciprocal, from the pulpit to the congregation. Testers did little to amplify the sounds coming from the congregation, which made whispering and murmuring—particularly in the farthest reaches, where the lowest status people sat—a dangerous, even seditious activity. Whisperers stole the minister's rightful audience. The congregants literally absorbed most of the direct signal with their bodies. High and steep ceilings were permissible, however, and whatever part of the signal was not absorbed would reverberate slightly as it faded off, a tone that would be perceived as a sharp ring to loud or staccato utterances.

Fig. 3.2 Interior of the Thomaskirche in Leipzig. Courtesy of AKG London.

The Thomaskirche in Leipzig illustrates the typical process of reforming a church's acoustics well. In the pre-Reformation church, the priest's voice would take a full eight seconds to fade away. Sometime in the mid-sixteenth century, the Thomaskirche was refitted for Lutheran services. The high ceilings were draped over to muffle reverberation. Galleries were added that further dampened resonance. The center of aural focus was moved away from the narrow side where the old altar can still be seen; it moved forward in the chancel and was stripped of its images and statues. Instead, a capsule-like pulpit was placed at the center of the widest side. The minister's voice was directly projected over the shorter distance to the congregants below and to those in the galleries rather than beginning as a reflected sound. The sounding board or "tester" above the minister's head and the wooden board immediately behind him served to amplify his voice and direct it onto the congregation. Longer echoes were muffled by the bodies of the audience and the drapes on the ceiling. Reverberation time was one-fifth that of the pre-Reformation church, taking only about 1.6 seconds to fade away. These "reformed" acoustics made it possible for one of the church's eighteenth-century cantors, Johann Sebastian Bach, to write intricate organ and voice music full of nuances that would have been lost in the old church. They also made it possible for congregants to comprehend the sermons and readings of the minister.[6]

Fig. 3.3 Globe Playhouse, London. Courtesy of Guildhall Library Corporation of London.

Reformed church acoustics also resembled those of open theaters, where again clarity of voice was more important than fullness. Compare the Thomaskirche to the old Globe Playhouse in London, for example. Both sought clarity by stacking up as many seats as possible at the closest distance from the aural center. Draping the ceilings of the church served the same purpose as the open air above the Globe: sound went up but never came back down as an echo. Bruce Smith calls the wooden stage a sounding board for the actors, making the comparison with the pulpit furnishings explicit. But Reformation churches differed from London theaters in ignoring vision, placing a number of benches facing away from the pulpit, though still within good auditory range.[7]

North American Houses of Worship

The acoustics of meetinghouses and churches reflected early American methods of constructing and maintaining social order both within a society and between the heavens and that society. Chesapeake churches, hewn closest to the Church of England, became the most hierarchical. Quaker meetinghouses began as the most egalitarian, but the years led inexorably to acoustic hierarchies, though Friends maintained a reciprocity that others lacked. Puritan meetinghouses also began with egalitarian acoustics, but in ordering their interior acoustic spaces they, like Chesapeake Anglicans, became more hierarchic and unidirectional, with sound and the authority to make it flowing from heaven through the pulpit and onto the audience. For it to flow back in any but sanctioned forms was a transgression and a threat to social order in both the Chesapeake and New England.

All the early houses of worship shared certain characteristics as a result of their common origin as well as circumstance. Unlike Roman Catholicism, worship was in English, so all three denominations sought a degree of acoustic clarity, although how much and for whom differed in important ways. Seventeenth-century churches and meetinghouses were usually of the "auditory" or "hall" type rather than the larger, cross-shaped, and more reverberant basilicas that tended to be the pre-Reformation urban norm in old England.[8] Communities began small and so it was with their churches and meetinghouses. They all started from scratch so Reformation acoustics could be built in rather than added on. Despite their commonalities, the differences show us much about the pluralistic construction of social order and semi-public space in seventeenth-century British North America.

Chesapeake Churches and Chapels

Jamestown's settlers carefully constructed the soundscape within the chapel that marked the center of their day-to-day existence. They were particularly proud of their cedar pulpit. While no mention is made of a canopy or tester, the pulpit was probably raised somewhat and placed near a reflective wall in the 1610 chapel.

By the early eighteenth century, American Anglican pulpits generally raised the minister above the audience to make him easier to hear. A sounding board or tester was usually part of the pulpit, particularly in larger churches. It would be either carved out of the same block as the pulpit or suspended overhead. It concentrated and reflected the minister's voice, amplifying it to make him seem louder than an untreated voice. At the bottom of the pulpit would be the "reading desk" from which the bulk of the service would be read out. At the ground level was another desk from which psalms and hymns were "lined out" from a book. From the reading desk the voice would reflect from the walls of the pulpit, giving the reader an acoustic position almost as prominent as when the minister spoke.[9]

Like their brethren across the Atlantic, Chesapeake Anglicans focused the sounds that came from positions of authority. The Jamestown chapel itself was two and a half times longer than wide, and the pulpit and communion table were most likely situated at the eastern end, particularly if there was a chancel screen. The chancel, at the east end of the chapel, was in the building rather than attached to it, which leads Dell Upton to think that the chapel was of the auditory type.[10] Sounds were projected along the length rather than the breadth of the chapel. Where one was seated was an indicator of social standing.[11] Those in the front heard more of the direct signal. Those in the back heard less of it, and thus more reverberation. This may have reinforced social assumptions. The people seated in the back from lower social standings were assumed more likely to be taken in by the power of the sounds rather than the articulated words. Although we know little about the height of the earliest Jamestown chapels, the brick church built there in the late seventeenth century had high, steep, reverberant ceilings.[12]

The congregation's job was to listen rather than see. As early as 1610, the Jamestown chapel may have attenuated the visual with a cedar chancel screen that blocked the view of the communion table. Chancel screens were the norm in later Chesapeake Anglican churches.[13] The primacy of listening over seeing was also reflected in seating. Rude benches probably gave way to box pews for many elite churchgoers sometime in the seventeenth century. Box pews had high walls to keep out drafts. Seating was on

Fig. 3.4 Second Bruton Parish Church, 1681–83, Middle Plantation (now Williamsburg), showing the high-ceilinged, narrow layout typical of seventeenth-century Chesapeake churches. Brick churches did not become the norm until the last quarter of the century. Others built in this style include the old brick church at Jamestown, traditionally said to have been built sometime between 1639 and 1647 but which Upton places in the 1680s, and the Newport Parish church, built about 1685. The drawing was made in 1702 by Franz Ludwig Michel, a Swiss traveler. Dell Upton, *Holy Things and Profane: Anglican Parish Churches in Colonial Virginia* (New York and Cambridge, Mass., 1986), 39. Photograph held by Colonial Williamsburg Society. Courtesy of Burger-bibliothek, Bern.

three or four sides of the box so that many auditors were facing away from the pulpit and reading desk or unable to see because of the walls. Poorer folk and, later, slaves found seats in single pews in the back or in the rear galleries. The box pews partially obstructed the views of those in back at ground level.

The congregants were supposed to make a "joyful noise unto the Lord" at the appropriately sanctioned times. The emphasis on noise rather than sound meant that articulation was not as important as volume. God knew

Fig. 3.5 An Anglican chapel from the Eastern Shore of the Chesapeake, built about 1771 (Christ Church in Broad Creek Hundred, near Laurel, Delaware). Although built late, it shows some of the design principles that were operating even in humbler Anglican churches of the early eighteenth century. Note, for example, the reflective wooden walls, the pulpit and tester, and the box pews. Courtesy of the Library of Congress.

the words; it was the sincerity and affect of the response that was important. Here the high ceilings played an important role. They strengthened the congregation's joyful noise, which, unlike the minister's voice, was not directed away from the ceiling by the sounding board or pulpit and would thus reverberate more, but at the expense of clarity. The vigorous ritual sounding of the congregation's set responses placed the community in its proper relation to itself and the heavens. The reader and the minister had to be clear because the congregation—unlike God and the speakers— could not be assumed to know all of what was being spoken from the desk and the pulpit. The congregation was supposed to be loud and powerful, and the high ceilings helped.

Chesapeake church acoustics shifted toward the end of the seventeenth century. Brick churches began to replace the smaller wooden chapels, though the dimensions remained the same. These in turn gave way to larger, more familiar churches of the eighteenth century.[14] The later churches were no longer at the center of Chesapeake daily life, replaced by the peculiar rhythms of the plantation. Less care was given to the placement of the pulpit for good hearing in plantation-era Chesapeake churches. Rather than placing the pulpit against a wall, it would be on a corner of the chancel with open air behind it. The testers themselves became more decorative and less functional, with fancier ones having no acoustically reflective surfaces at all.[15] Eighteenth-century Anglican ministers often read their sermons in a nearly inaudible mumble to avoid being associated with their more evangelical brethren. High ceilings and the lengthwise orientation of the church remained. Cross-shaped churches, usually with chancel screens, began to be built again. All these tendencies reinforced a loss of interest in acoustic clarity, as the importance of the Chesapeake's visible world waxed in the eighteenth century.

New England Meetinghouses

In seventeenth-century New England meetinghouses, acoustic qualities weighed heavily against visual factors. There were no statues, altars, images, or paintings—all of which would have been considered graven. The eye would be immediately drawn to the pulpit, a large, centrally located elevated capsule, usually supported against one of the long walls if the building were rectangular. This was the place where ministers expounded and explicated the word, the Bible. Like bells, an inordinate effort went into the procuring of properly constructed pulpits in seventeenth-century New England. Some were transported from other areas at considerable expense, and if the work of a particularly valued craftsman could not be had, his style might be carefully copied. Although the pulpit would be the visual center of the room, this factor was subordinate to the consideration of acoustic properties in selecting one, particularly the sounding board carefully angled over it.[16]

New England meetinghouses, like the chapel in Jamestown, were at the auditory center of the town. Unlike the Chesapeake, however, meetinghouses remained at the center of New England town soundscapes throughout the seventeenth century. This was explicitly stated in the Massachusetts Colony Records, where one order from 1635 stated that "noe dwelling house shall be builte above half a myle from the meeting house" without express permission from the colonial court. Towns were laid out in six-mile squares and meetinghouses were supposed to be at the center. Dwellings were not evenly scattered through the remaining area. They were clustered nearer the center with a complicated system of lots on the outside perime-

Table 3.1. Average Dimensions of Seventeenth-Century New England Meetinghouses

	Shape			Avg. Dimensions (feet)			Ratio	Avg. Area	Avg. Vol
	Square	Rect.	Unk.	L	W	H	L/W	sq. ft.	Cu. Ft.
1631–42	3	2	31	37.2	35.0	12.0	1.1	1,302.0	15,624.0
1643–60	5	11	27	36.0	30.0	11.9	1.2	1,080.0	12,892.5
1661–80	5	22	34	40.0	33.0	15.6	1.2	1,321.5	20,648.4
1681–1700	9	26	24	42.1	35.9	17.3	1.2	1,510.3	26,190.1
Total	22	61	116	38.8	33.5	14.2	1.2	1,299.8	18,490.7

Source: Data compiled from Donnelly, *New England Meeting Houses of the Seventeenth Century,* 121–30.

ter, where no more than a crude day shelter was supposed to exist. Thus, every home was expected to be within a mile or so of the meetinghouse.[17] This was well within earshot of a small bell, a drummer, or conch shell.

Seventeenth-century Puritan meetinghouses in New England were proportionally wider than their Chesapeake Anglican counterparts. While the typical ratio for an Anglican church during this time was about two and a half times long as wide, New England meetinghouses were often square or nearly so, with very little change in the layout of floor space until the eighteenth century. As populations grew, meetinghouses grew larger vertically. Many of the later-seventeenth-century meetinghouses added on a gallery. In this way, the principles of the auditory church could be stretched to fit the greatest number of listeners at the least distance from the pulpit. Basically, meetinghouses grew larger in the seventeenth century by stacking people up in two levels surrounding the raised pulpit. Only in the eighteenth century did Congregational meetings adopt longer, narrower church layouts.

None of the earliest meetinghouses survive. Descriptions of repairs, additions, and remodeling point toward a squarish shape as the norm. Rectangular buildings tended to be not much longer than wide, with the pulpit usually set up on one of the long walls. Galleries were often added a few years later rather than at the time of building.

Until the 1680s the information about meetinghouses is too sketchy to make more than conjectural attempts at reconstruction. Three seventeenth-century New England meetinghouses have left enough information for a discussion of their acoustics, however. These three buildings—Hingham's "Old Ship," Plymouth's second meetinghouse, and Deerfield's third meetinghouse—were, with exceptions noted, fairly representative of New England patterns in the seventeenth century.

"Old Ship," the meetinghouse built in Hingham in 1681, is the only surviving New England meetinghouse from the seventeenth century. At seventy-three feet long and fifty-five feet wide today, it is quite a bit bigger than when first built. Nonetheless, records of its upkeep and remodeling are thorough enough that we can get a fair picture of its inside and outside in the late seventeenth century. In 1681, it measured fifty-five feet in length and forty-five feet across.

"Old Ship's" full volume needs to be taken into account in order to understand its acoustics. It had a high vaulted ceiling said to look like the inverted hull of a ship from the inside. Three of the galleries were built into the original plan, running the length on one side and along the width of both sides. The two-story pulpit rose up halfway between the floor and the galleries against the wall of the other long side. The tester atop the pulpit was positioned a little above the galleries, directing the minister's voice across to them and also downward to the floor-level pews. The tester prevented the minister's voice from being lost in the vaults of the high, open ceiling. The pulpit was less than twenty feet from the furthest front edge of the long wall gallery, and as near as six feet to the short wall galleries.

The meetinghouse's acoustics need to be considered while full of people and without amplification introducing distortions. Sound coming from the pulpit would be absorbed in much the same way that the audience at the Globe Playhouse absorbed sound from the stage. The resemblance to a theater was more than chance. The Protestant theologian Johann Valentin Andreae (1586–1684) proposed an ideal house of worship to be situated at the center of town and built in a way that "the ears of

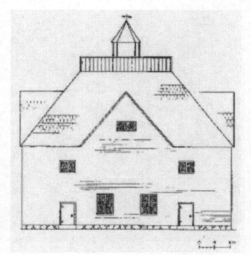

Fig. 3.6 "Old Ship," Hingham's second meetinghouse, as it is conjectured to have appeared in 1681. Drawing by Marian C. Donnelly, in Donnelly, *New England Meeting Houses of the Seventeenth Century,* 76. Courtesy of Wesleyan University Press.

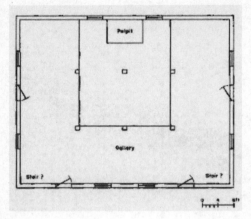

Fig. 3.7 Plan of galleries and pulpit, "Old Ship," 1681. Drawing by Marian C. Donnelly, in Donnelly, *New England Meeting Houses of the Seventeenth Century,* 78. Courtesy of Wesleyan University Press.

all may be equally distant on all sides from the voice of the speaker." The imaginary building was also meant to be used for religious drama, and in the words of Marian Donnelly, arguing for the influence of Andreae's design on New England architecture, "a relation to theater design seems clear."[18]

Only the sharpest of the minister's utterances would give the long reverberant ringing from the high ceiling. Why then build such a huge vault? The congregants' response would carry up to the ceiling and reverberate extensively, creating the perception of a clearly articulate minister's voice joined by a less clear but sonically fortified response. "Old Ship" was a specialized acoustic instrument, one designed to be heard from the inside and

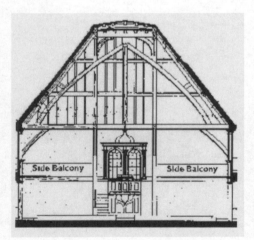

Fig. 3.8 Cross-section of Old Ship meetinghouse, 1681. Note the height of the pulpit and tester, as well as the large ceiling area. Adapted by author from drawing by Murray P. Corse in Corse, "The Old Ship Meeting House in Hingham, Mass.," *Old Time New England* 21, no. 1 (1930). Used by permission of the Society for the Preservation of New England Antiquities.

Fig. 3.9 Plymouth Meeting House. The sketch in the top right corner probably depicts the 1683 meetinghouse. The main sketch is the same building with two of its cross gables removed and the length extended. Courtesy of the Pilgrim Society, Plymouth, Mass.

to clearly define the relationship of congregants to minister and all worshipers to the heavens, with the heavens conceived not as a visual space above, but as the invisible, present world within, a world that could be bridged by the auditory.

An intriguing diagram of the Plymouth meetinghouse of 1683 points toward a long rectangular building rather than a square one. From town records, however, it is clear that the rectangular building was created by taking two of the four gables down from the original, squarer building and then adding on to the building on each side. When finished, the building had a much more symmetrical facade with two gables rather than four, prefiguring Georgian lines that would come to dominate the housing patterns of the elite in the eighteenth century.[19]

Enlargements of meetinghouses did not always go from square to rectangular. The first meetinghouse at Framingham began as a rectangle but was enlarged to square it off in 1715.[20] In no instance was a seventeenth-century meetinghouse made more than twice as long as its width, the norm for the Chesapeake. This allowed Congregationalists to maintain the au-

Fig. 3.10 Deerfield's third meeting-house, c. 1694. Dudley Woodbridge Diary, October 1–10, 1798. Courtesy of the Massachusetts Historical Society.

ditory-style hall as a viable type well into the first half of the eighteenth century.

A sketch of the third meetinghouse, in Deerfield, portrays a tall, hipped-roof building. It had a belfry on top. The square shape and the two-story structure mark it as New England's variant of the "auditory" meeting-house, equipped with galleries and a tall pulpit with a tester. The high ceiling acted acoustically as the vault in "Old Ship" did. A tester on the pulpit kept the minister's voice clear and loud, while the vault above would reverberate the congregation's collective voice, reinforcing their ties with the invisible world.

Acoustics influenced the seating of the congregation, too. Robert J. Dinkin has mapped out how Puritans enacted their social orders in meetinghouse seating arrangements, arguing that they were seated in ways that visually reinforced differences in status and wealth within the community. This system of deferential seating, he says, was not challenged until after the Revolution except by those who felt they were not seated well enough. In order to smooth over tensions in this regard, some seats which did not appear to be visually well located were "dignified" as high-status seats. Dinkin does not offer a reason why such status-conscious people would accept such an arbitrary solution.[21]

If the acoustics of deference is considered, two things become apparent. First, deference operated in different ways in the audible world and in the visible; and second, the "dignifying" of seats was not arbitrary. The visible status of Puritans individually and collectively was unsettled and contested throughout the seventeenth century. Collectively they thought of themselves as a "city on a hill" at first, but then later as a "saving remnant" somehow lost on their errand into the wilderness. Individually, the question of just who was among the elect, the "visible saints," nearly divided the movement in New England in the 1660s. Acoustic order was much clearer, and perhaps in some ways a cohesive unifying force, not because of some inherent properties of "orality" but simply because it was already worked out and agreed upon in the seventeenth century. In the meetinghouse, the minister articulated the voice of God for the congregants, whose auditory task was first and foremost to hear clearly and respond in a set way, loudly and with their hearts. Transgressions were seditious. Anne Hutchinson, Mary Dyer, those accused of witchcraft both at Salem and before: these were disorderly speakers, and in the words of Hutchinson's accusers, their "voluble tongues" were a good part of what made them threatening.[22]

The acoustic environment of the meetinghouse was expected to provide all the godly and even the sinners with the ability to hear the minister clearly. When four members of Haverhill's congregation complained about their seats, they petitioned on the basis of audibility rather than visibility, saying they were "obliged to sit squeezed on the stairs where we cannot hear the minister and so get little good from his preaching, though we endeavor to ever so much."[23] One congregant proposed that sinners ought to get the best seats because they were most in need of hearing. Dinkin calls the seats that were dignified "less desirable" and says that it was a somewhat arbitrary ruling to make certain seats "equal in dignity to the better seats nearest the pulpit." He then discusses the "dignifying of seats" in Marlborough, noting that the front seats in the gallery were next in dignity to the second seats below and that the front seats of the side gallery were next in dignity to the third seats.[24] If the "dignity" of a seat is thought of as both the visible location and the auditory location of the seats, the decisions become clear. The third seat on the floor may have been farther from the pulpit than the front of the side gallery. And the front row of the front gallery, while out of visible range from many of the congregants below, was in an excellent auditory location, being directly in front of the tester. Obviously visibility was a status marker for the seated, but audibility may have been on the minds of the seating committee and the minister. This way of thinking was undergoing change at the end of the seventeenth century, and perhaps it is best to take the most egalitarian of early Anglo-Americans, the Quakers, to see how.

Quaker Meetinghouses

Many early Quaker meetinghouses were square or nearly so, hexagonal or octagonal. The earliest meetinghouse for which we have evidence was

Fig. 3.11 Burlington's first Quaker meetinghouse, 1683. Note the hexagonal shape, and the cupola, which may have housed a bell or a drummer. Original oil painting, Quaker Collection, Haverford College. Used by permission.

hexagonal, with a hexagonal roof. This was the first Burlington meetinghouse, built in 1683. It hosted the yearly meeting for the East and West Jersey and Philadelphia every other year from 1683 to the mid-eighteenth century.

The Burlington meetinghouse's shape acoustically instituted Quaker notions of egalitarianism. From the inside of the building, the ceiling panels acted as a set of six sounding boards, equally amplifying voices originating anywhere in the room. While the Burlington Meetinghouse is gone, other octagonal and hexagonal buildings are still extant. From one of these, the hexagonal Quaker schoolhouse in Birmingham, Pennsylvania, we can reconstruct the acoustics of an earlier time. The acoustics are very crisp. The decay time for reverberation is quick, much less than a second, because of the relatively shallow pitch of the roof. The walls and roof echo the voice directly back into the audience rather than reverberating like a steeper roof. Because the delay time of the echo is so short, the echo is experienced as a fullness to the voice that does not impede clarity. A round

Fig. 3.12 Birmingham, Pennsylvania, Quaker school house. Photograph by author.

room or a dome might seem even more equitable, and in fact earlier Protestant architectural reformers thought it to be the best shape, but acoustically a round room or a dome would direct sound too narrowly, conveying a voice from one point to another without diffusing it. This would create "listening spots" rather than making voices clear and audible to all. The flat surfaces of square, hexagonal, or octagonal buildings would be just uneven enough to refract the voice rather than concentrating it, so these shapes were preferred over circular walls and ceilings.

The efficacy of these rooms in conveying the voice is attested in two ways. The Birmingham schoolhouse has hooks in the ceiling for draperies in order to dampen sound. Apparently, the combined voices of a roomful of children were amplified so well that they had to be countered to make the room more conducive to learning. Also, shape-note singers today often seek out Quaker hexagonal and octagonal rooms for their singing. Shape-note singing is usually done with a roomful of participants—there is no audience per se, just as everyone is a potential participant at a Quaker meeting. The emphasis in shape-note singing is on the mid-range frequencies that mark a clearly articulated voice, the same frequencies emphasized in Quaker plain styles of speech.[25]

Quakers were not alone in their use of hexagonal meetinghouses. The eighteenth century witnessed the construction of several others among

Fig. 3.13 Birmingham Quaker school house, interior. Photograph by author.

Wesleyan, Congregationalist, and Dutch reformed congregations, but the early and most emblematic use was by the Quakers. The old folk saying that attributed the hexagonal design to the notion that there was no corner for the devil to hide is perhaps a way to think about reverberation. Other early Quaker meetinghouses tended to be square and have galleries, much like the early Puritan meetinghouses, but with no pulpit.[26]

The soundscape of the room was critical to Quaker worship, and its delicate interplay of silence and speech rested uneasily in a bustling urban center like Philadelphia. Maintaining silence was often a challenge. The "Great Meeting House" at Second and Market Street in Philadelphia was driven into disuse in the eighteenth century because the street noise became too much. Iron hooves and cartwheels on cobblestones and pavement made a tremendous racket, much louder than present-day automobiles. Even with specially designed roadways with cobblestones in the middle to give the horse traction and smooth paving stones for the wheels, the noise in an urban environment could quickly overwhelm a meeting. Noisy children and barking dogs posed another set of problems, which meetinghouses tried to solve with gates and gatekeepers.[27]

Inside the room, careful attention was paid to acoustics above and beyond the shape of the room. Since everyone was in theory a potential

Fig. 3.14 Bank Street meetinghouse, ca. 1683. Philadelphia Meeting Houses, 911 A-F Box 1, Quaker Collections, Haverford College. Used by permission.

Fig. 3.15 Reconstruction of seventeenth-century cobblestone and pavement roadway, Elphreth's Alley, Philadelphia. Photograph by author.

Fig. 3.16 One of the sounding boards above the facing benches in the Arch Street Meeting-house, Philadelphia. Notice the width and height of the board, both of which serve to collect sounds from the whole room, including the galleries, in addition to projecting sounds. Photograph by author.

preacher, the minister-audience dichotomy of other denominations would not work. Acoustics had to be clear and sharp everywhere, for all speakers and hearers alike. In practice, however, certain concessions to hierarchy were made as the meetings grew larger. A set of "facing benches" was set above the other seats and often had a curved wall behind them to both collect and project voices. Although acoustically favored, the seats were sometimes called a gallery to make them seem less elite. The acoustic differences between these seats and the New England or Chesapeake minister's pulpit were significant. Obviously, more people were seated in these seats than in the pulpit. But more important, there was an element of reciprocity to the facing benches. Other denominations focused the sounds of the minister's voice sharply and directed it toward the audience. No care was taken to clarify the audience's joyful noise, and no effort was made to make the minister's pulpit a favored place for *listening*. The size of the sounding boards behind the facing benches did just that, though, collecting and amplifying sounds coming in from the room as well as projecting the voices of the "elders" who sat there. The importance of being able to listen was underscored in 1763 when a new stairway impaired the acoustics of the High Street Meeting in Philadelphia, and the meeting found it necessary to "fix up a suitable board for the conveyance of the voice when Friends are concerned in public testimony."[28]

During the eighteenth century all denominations introduced more hierarchy and emphasized visual over auditory elements in reinforcing social order. Sounding boards fell into disrepair, with some ministers fearing injury so much as to have them removed from the pulpit. Slanted, acoustically effective testers were reset parallel to the floor, which improved them visually but defeated their sonic purpose. Churches were introduced into Congregationalist architecture, so that the Chesapeake and New England acoustic soundways converged. Quakers maintained some amount of reciprocity but sacrificed the primitive egalitarianism of the early meetings to a de facto system of elders who were expected to speak more often and non-elders who were expected to remain for the most part silent.

Perhaps because of its very nature, sound remained difficult to regulate and order. Rather than improving regulations, diminishing the importance of sound was ultimately how early Americans made their soundscapes more manageable. Before they did so, however, they spent a tremendous amount of effort on governing each other's tongues, with ever-diminishing returns. The next chapter explores vocal soundways without yet considering language.

CHAPTER FOUR

On the Rant

I used to play in a band—a punk band that dabbled in reggae in the early 1980s, when such things were yet transgressive: loud, angry music with words, such as they were, barked, howled, roared, and yelled. They were rants as much as songs. We placed ourselves outside any "civil" conversation. No efforts were made to exclude. With few exceptions anyone was welcome to our little world, but only a certain fringe of society was attracted to it. We fought incessantly, and played, in all senses of the word. We were an unstable referent, with the line between audience and band often dissolved or at least blurred beyond meaning. Our imagined location was on the edge, the margins, outside looking in and commenting on what we saw, felt, and heard in ways only the disaffected could adopt. Yet we were at our own center, if sometimes only by dint of volume controls. At the same time, what it meant to be a center was constantly interrogated, as our group—both within and outside the band—poked fun at the pomp of more serious claimants.

This chapter concerns the sounds of voice that fall outside the realm of what can be captured by an alphabet and rendered visible. These are the paralinguistic aspects of speech (such as tone, volume, cadence, and pitch) as well as vocables (nonlinguistic vocalizations such as groans, howls, sighs, and roars). Early Americans attributed great power to such sounds, considering them tangible forces. Rather than considering a world of powerful sounds as superstitious or magical, the present chapter delves into what sorts of civil and uncivil spaces early Americans constructed from paralinguistic and vocable sounds.

Clamor, discourse, humming, murmurs, muttering, railing, rants, roaring, swearing, and whispers: Each can be mapped along an earthly plane locating it in respect to civil society. Each can also be located along a spiritual plane. This three-dimensional aural space defined orderly—and disorderly—societies in seventeenth-century English North America. After mapping some of that space, this chapter uses a single case study—that of the "ranting Quakers"—to explore the centers and edges of three such spaces, each overlapping the other: one British, one Puritan, and one Quaker.

Such worlds had boundaries between inside and outside. On the horizontal plane, the boundaries were the edges of civil society, beyond which fell the savage or the wild.[1] On the vertical plane, the boundaries were between the visible and invisible worlds. Sounds crossed these thresholds and made the borders more complex and semi-porous—and also more dangerous. The wild and the other-worldly mixed to produce demoniacal aspects of the invisible world. The civil and the other-worldly optimally came together as a tenuous connection to the divine that depended very much on social order and the careful interpretation of cultural expression. This in part explains the colonists' massive efforts to control traffic in speech, particularly where matters of religion and government were concerned, and especially at the boundaries.[2]

Some nonverbal aspects of vocalization were on the material plane. "Clamors" caused disaster in Virginia, according to John Smith. Some "projecting, verball, and idle contemmplators" among the Jamestown colonists would say anything to those on the supply ships in order to get a little extra food, passage back to England, or else a name at home by providing wild stories about the state of things in the colony. "Thus from the clamors, and the ignorance of false informers, are sprung those disasters that sprung in Virginia: and our ingenious verbalists were no less a plague to us in Virginia then the Locusts were to the Egyptians." Smith claimed that clamors were a worse problem than illness, short supplies, or Indian attacks combined.[3]

By labeling the speech of the disaffected as "clamor," Smith was able to discount the content of the respective criticisms. Clamors, then, were unwelcome critiques. In a rank order society such as Virginia, critiques from below could be dismissed for their content and treated as unruly—even seditious—acts in which the disruptive nonverbal elements were given precedence over the linguistic content. By this line of thinking, it was not the reasoning contained in them that had effects, but the sound itself.

The opposite of clamor was discourse. Discourse was conversation operating within social parameters. As a verb, it held the meaning now denoted by "discuss." Discourse took place out loud and invisibly. It was the negotiation of meaning via an orderly succession of linguistic sounds. It was the very process of reason and rationality. These sound-centered usages of the term became obsolete by the end of the eighteenth century, but they were the main senses of the term until the late seventeenth century. As the variants concerned made a long slow fade, another meaning of *discourse* began to gather momentum, becoming primary in the eighteenth century. This was a reasoned structure of ideas, not necessarily taking place within the context of a conversation. A discourse became any rational structure of thoughts, disconnected from the medium of sound and independent of conversation. To engage in conversation became merely to discuss, with the connotation of reasoning lost. Sound became detached from the process of reason only in the eighteenth century. In the seventeenth, it was integrally connected to the processes of thinking, whether rationally or irrationally.[4]

Insubordination and dissent manifested itself in quiet nonverbal vocalizations as well as loud ones. The meanings of these sounds were understood throughout the English portions of the colonies. Murmuring, grumbling, and whispering were the most common of the quieter problems. Left unattended they could lead to ranting anarchy.

Murmuring could threaten the very survival of a community. A series of "devilish disquiets" plagued the unity of the frail little colony of shipwrecked sailors and colonists on Bermuda in 1609. Rumors that Virginia was nothing but work and wretchedness started "first among the seamen, who in time had fastened onto them (by false baits) many of our landmen likewise. . . . This, thus preached and published each to other, . . . begat such a murmur" that it threatened the company's ability to effect an escape. A minister's clerk named Stephen Hopkins made "substantial arguments both civil and divine (scripture falsely quoted)" that Gates's authority as governor ceased with the shipwreck, and that "they were all then freed from the government of any man." The governor declared that Hopkins and his followers were guilty of "murmuring and mutiny" and sentenced Hopkins to death, although he "made so much moan" that he was granted a reprieve.[5]

Murmuring was not always put down. It sometimes made elites reconsider their positions, as when the enforcement of royally proclaimed trade restrictions lapsed in the face of murmuring from small planters in Virginia in 1626. Sometimes murmuring was minded by well-ordered Christians: even among the apostles, wrote Quaker Robert Barclay, "there was a Murmuring that some Widows were neglected in the daily Ministration" that they had to redress.[6]

Increase Mather, reflecting on the causes of Metacom's war, asserted that providence had made it so harsh because "the people" were "Full of murmurings and unreasonable Rage against the enemy." Seventeen years later, his son concurred. "We have been a most Murmuring Generation," wrote young Cotton Mather in 1693. The outbreak of what the younger Mather perceived to be witchcraft was to his mind the result of the murmurs of discontent that had passed through New England in waves. Murmuring meant that people were unsettled and left room for the devil to gain entrance to the society.[7]

Murmurs could be fed with rumors to create a threat to civil order. Robert Beverly wrote in 1705 that the "poor People" of Virginia had become uneasy due to religious intolerance, poor economic conditions, and inequitable distribution of wealth. "Their Murmurings were watch'd and fed" by "mutinous and rebellious Oliverian soldiers" until a servants' rebellion broke out.[8]

At his commencement (with a Master of Arts) from Trinity College in Dublin, Increase Mather "refused to comply with the usual Formalities of Hoods, caps etc." The school, being run by Presbyterians, was somewhat annoyed with him, but "Many of the scholars were so farr pleased" with Mather's resistance that "they did publickly Hum me." Mather claims he was greatly surprised when he "heard the scholars (many of whom I knew not) begin their Humming." Public humming was a way for members of a crowd to safely express approval or occasionally disapproval—perhaps only because no one could see their lips move.[9]

As with the earthly dimension, one's location in the spiritual plane was marked at any given moment by a person's vocalizations, linguistic and otherwise. Through births or deaths, early Anglo-Americans panted, moaned, and groaned to create a liminal space at the threshold of the physical world. For example, Anglo-Americans called childbirth a woman's groaning time or a groaning. A woman in labor ushered the newborn's arrival into this world. This was a time of high mortality for both infant and mother. The woman's groaning marked the liminality, the in-between-two-worlds quality of the process. Kathleen M. Brown notes that women's oaths—worth little in other contexts—were considered infallible when naming the father during childbirth in paternity cases. When a woman was in her groaning time, Anglo-Americans believed her voice was in touch with the divine and she could not lie. Groanings had their own rituals and practices. There was groaning pie and groaning bread, even groaning beer. Once finished, the woman recovered in an oversized groaning chair next to the bed.[10]

In Puritan New England, groaning was also a mark of a heartfelt attempt to converse with the divine. Recovering from the "Ephialtes"—nightmares—in 1671, Increase Mather (who feared them as a portent of oncoming madness) "poured out my [his] Heart with many Tears and groans before the Lord." The process of being saved or "convinced" of one's faith involved a vocal but nonverbal entreaty to the divine. According to Mather, the would-be saint "groan[ed] with inward tumult, insufferable throbs and sighs." If successful, "the new birth has pangs attending of it that make the convert cry, lift up his voice and cry aloud for mercy." His son, Cotton, advised would-be converts thus: "Oh! Make thy Moans and thy Groans, even the Groans of a deadly wounded Man" unto God. Eventually this moaning would effect the appropriate fear that was an initial step on the road to conversion. Ineffectual or weak prayers, those with little chance of crossing the threshold to the divine, were said to be lisped. In 1728, Salem judge and former deputy governor John Danforth wrote about the pathetic response of churchgoers to the warnings of Providence, particularly the sounds of a recent earthquake. According to Danforth, the congregants did not even lisp the echoes of a response to the earthquake. Then he turned toward the state, asking again if there was no lisped response to the "shrieks" that "loudly groan" of King George's death.[11]

Extralinguistic and paralinguistic vocalizations were thought to be somehow more immediate expressions of the inner self than linguistic voicings. Perhaps people believed that such sounds were connected to their God in a way prior to rationality. Language and thought were accretions to be stripped away in communicating with the other world, signifiers of human grandiosity and weakness that meant nothing when confronted with the divine.

Possession

The sounds of possession fed into both the spiritual and the material plane. For seventeenth-century Anglo-Americans, everyone was possessed. They understood good Christians to be properly possessed by God. People who were otherwise possessed, whether Indians in the wilderness or Ranters on the edges of civil society, were thought to be possessed by self-will or demonic forces. Properly possessed servants, slaves, wives, and children belonged to male masters. The masters and their families belonged to congregations, which in turn belonged to God, as interpreted by ministers.

Seventeenth-century Anglo-Americans knew that all sounds ultimately had articulate sources. For humans, articulation is a function of language alone. If one could locate a visible articulator, then the source was found. But the human voice could be inarticulate, or it could have an inarticulate dimension to it. These sounds indicated possession of the utterer by some invisible will, either the person's hidden intents, or diabolical forces, or even Providence. This trait was the principle behind the power of "heated" speech, namely that the utterer was not in proper possession of the sounds uttered. Powerful vocalizations like psalm-singing placed the utterer solidly in the center of things as they should be. For Puritans and others, a well-sung psalm or hymn could allow a feeling of immanent grace to overcome one's being, a feeling of belonging to the divine. Other signs of possession were not so good.

Increase Mather was able to list six general symptoms of possession by demonic forces. Three of them were auditory: first, that the possessed would "reveal secret things" that could not have been known otherwise; second, that they would speak in "strange languages." The third was "uttering words without making use of the organs of speech." All these utterances were outside the normal articulatory powers of the utterer.[12]

In 1662, Ann Cole, "accounted a person of real Piety and Integrity" by her Hartford ministers, "was taken with very strange Fits, whereby her tongue was improved by a d'mon to express things which she her self knewe nothing of." The demons supposedly said through Cole's mouth, "Let us confound her language that she may tell no more tales." She then began muttering incoherently. But then "the Discourse passed into a Dutch-tone," and she began to mutter in English. The "Reverend Mr. Stone (then Teacher of the Church at Hartford)" thought it impossible that one as ignorant of the Dutch language as Cole could "so exactly imitate the Dutch-tone in the pronunciation of English." Mather makes much of this, because speaking in foreign languages that one did not know was a sure sign of possession. Mather wrote that Cole's "Malady was not a meer natural Disease" because "the *tone* of the Discourse would sometimes be after a Language unknown to her." The sound of her speech was a sufficient marker of possession, the actual content of the language remaining English. A lower-class Dutch woman implicated by Cole's Dutch-toned English was indicted and, upon hearing the charges that Ann Cole's tongue had passed, confessed to them all. She was in league with the devil, the poor woman admitted, and though she had signed nothing yet, the devil "had frequently the carnal knowledge of her body." She and her husband (who never confessed) were executed. Others accused of being responsible for Cole's possession fled.[13]

The case of Elizabeth Knapp of Groton, a "thing which caused a noise in the countrey" in 1671, serves to further illustrate some of the extra-lingual and paralingual features of possession. Knapp "was taken after a very strange manner, sometimes weeping, sometimes laughing, sometimes roaring hideously" for days at a time over the course of a three-month possession. This "strange manner" that clearly marked the otherness of what possessed her was wholly constituted of paralinguistic or extralinguistic vo-

calizations. At one point during her possession, a demon supposedly began to speak through her without the motion of her lips, again signifying
the "otherness" of her voice. Mather notes that "she was thought to be under bodily possession: Her uttering many things (some of which were
Diabolical Railings) without using the Organs of speech." Her minister,
Samuel Willard, claimed that "Her Tongue was drawn out of her mouth to
an extraordinary length, and now a D'mon began manifestly to speak in
her." Accordingly, she began to produce words without moving her lips,
even the labial sounds [p], [b], [f], [v], and [m], which require the lips to
move. Other times "words were spoken, seeming to proceed out of her
throat, when her Mouth was shut." Still other times her mouth was open,
but she allegedly did not use her tongue or vocal chords to speak. What
was said was notable partly for its content, but more so for its tone, being
"chiefly Railings and Revilings of Mr Willard." A series of blasphemies then
left her speechless for a time. She was finally cured of these sounds made
through her by confessing her sins out loud.[14]

In 1679, William Morse's son, of Newbury, Connecticut, was possessed.
During that time he supposedly "barked like a Dog, and clock't like an
Hen," after which he was unable to speak for a time, his tongue hanging
from his mouth. His parents tried to take him to a neighbor's house in the
hopes he would do better there. On the way "he made a grievous hallowing," threw a rock at the neighbor's maid, and began eating ashes, which
ended that plan. Back at home, he fell into another swoon and upon coming to, "roared terribly." They prayed for relief, and the sounds of the
devil—which had only barely made it to language at all—were defeated.
With a "a mournful Note" the ersatz devil cried out in a nonstandard variety of English: "Alas! Alas! me knock no more! me knock no more!" after
which there was finally quiet. Perhaps this was a construction of a West
African pidgin English. The devil was often construed as a "black man,"
and in 1700, Cotton Mather would go so far as to describe a devil supposedly seen in the possession case of Margaret Rule as the "Black Master" of
his minions. This devil would, continued Mather, "strike them and kick
them, like an overseer of so many Negro's [*sic*]."[15] The period from the
1670s through 1700 was one of anxiety over possession, as thousands of
possessed—for what better way to name enslavement—black Africans increasingly replaced the temporarily possessed indentured workers in the
southern colonies and in the ports of the northern ones. The connection
is further underscored by the fact that Rule could only consume rum, a
product of the slave trade.

Sometimes the battle for possession of oneself took place on the earthly
plane even for Anglo-Americans. In 1689, young John Gyles was captured

by the Cape Sable (Micmac) Indians and held as a slave, after which he was sold to the French. For several years, Gyles heard the vocalizations of the Micmacs and the French but did not know the languages. He did not possess a means of representing himself through language, although he was constantly exposed to its sounds. In both cases Gyles came to possess the language, and to some extent he possessed (or grasped) the culture of the "other" that possessed him. Once "redeemed"—that is, restored to the civil ownership of his native New England—he turned these possessions into a tidy career as a translator that banked on the knowledge of "others," both European and non-European, that he had obtained.[16]

Puritans were not alone in relating sound with proper possession. The Society of Friends sometimes "disowned" a member. To be dispossessed in this way left one open to possession by other forces. Sometimes these disownments were over what and how someone said something, as in the Keithian controversy at the turn of the eighteenth century, where George Keith was disowned in part for his tone in addressing other Quaker leaders. Quakers on the western frontier of Virginia continued to disown members who failed to lead "a sober and orderly life and conversation" right through the eighteenth century. Disowning created a silence around the disowned that amounted to a sort of social death. This withholding of acknowledgment of the disowned, this deadly silence, befell swearers and gossips, those who were "so far transported with passion as to utter some profane expressions," those who played music, those who answered with their names at military musters, and oathtakers, among others, and must have been a difficult sentence to bear in a small, closed community based on consensus, as were many Quaker meetings.[17]

Ranters

Somewhere between possessed and clamorous, a shadowy movement coming out of Long Island, known to us alternately as "singing Quakers" and "Ranters," posed a potent threat to social order in seventeenth-century British North America. In writing, Quakers carefully represented the paralinguistic and vocable aspects of how these "singing Quakers" sounded, holding them up to a set of implicit norms about civil behavior and finding them dangerously on the edges. In contrast, New England Puritans—led by Increase Mather and, later, Cotton Mather—constructed the singing Quakers as residing beyond the pale of civil society: their singing was never associated with words. It was an otherworldly possessed vocable sound. The Mathers constructed these Quakers, and by implication, all Quakers, in much the same terms that they used to write about Native Americans. They had their Quakers howling, roaring, and singing joy, but seldom saying anything. Pennsylvania Quakers, most prominent among them George Keith, countered Increase Mather by writing that the singing Quakers were "ranting" rather than making nonlinguistic sounds.[18] While "howling" and "roaring" placed the utterer outside civil society, "ranting" placed the utterer on the margins, with the paralinguistic vying with the linguistic for possession of the speaker's voice.

The civil societies that early Americans marked and maintained through their attention to paralinguistic and vocable sounds were not primarily visual spaces in the sense used in Jürgen Habermas's notions of public and private spheres. They were auditory fields that marked location and boundaries differently. In these auditory domains, civil power was conceived as the possessing of an audience. Paying attention to nonverbal vocalizations allows us to listen in on a world before the distinct conjuncture of print and tavern culture coalesced into a critical space from which emerged civil society and the public sphere. T. H. Breen has warned against projecting Habermas's notion of a public sphere onto periods before 1750 or so.[19] Surely, even if there was no public sphere, there was some form of civil society, a normative arena where "the people" were constructed together in public. In a world where state and religion were integral to each other, even as that relationship was splitting at the seams, Habermas's notion of civil society as a place outside the state will not hold. Attending to vocalizations in the same way that early Americans attended to them yields a civil society different from Habermas's in many ways. It was a world in which sound played an important role in marking and maintaining the limits of civil society. People sought a public hearing rather than a space in the public sphere.

The controversy over the so-called singing Quakers in the late seventeenth century sparked a debate over who was in and who was out of British civil society in the colonies. By the end of it, complex new social identities, not yet consciously American but no longer British, had begun to show through. Understanding them provides new insights into issues of plural American identities. Rather than a singular whiggish vision of America emerging in the mid-eighteenth century to form a nation from people who had previously been English men and women, we find here multiple, contentious visions of North American identity. Each was claiming to be the voice of the center, but all were hopelessly on the margins of empire, not realizing, to borrow Perry Miller's evocative phrasing, that they had been "left alone with America."[20]

In 1684, Increase Mather wrote of "the blasting rebukes of providence upon the late Singing and Dancing Quakers." According to Mather, the chief culprits of this were "three mad Quakers, called Thomas Case's Crew." One was a man named Jonathan Dunham of Plymouth. The second was a married women who was "following him up and down against her husband's consent." The third, who was the leader, was Mary Ross, a young woman from Boston.[21] Sometime in 1681, these three met up with Samuel Banks—"the most Blasphemous Villain, that ever was known in these parts"—and some others of like mind in Southhold, Long Island, and set to work on the senses, mind, and soul of Thomas Harris, a young Boston merchant in Southhold peddling his goods.

> They all got about him, and fell a Dancing and Singing, according to their Diabolical manner. After some time, the said Harris began to act like them, and to Dance, and sing, and to speak of extraordinary raptures of joy and to cry out upon all others as Devils, that were not of their religion, which also they do frequently: when the said Harris manifested these signs of Conversion, as they accounted it, they solemnly accepted him as one of their company; and Banks or Denham (for I have forgotten which of the two) gave him this promise, that hence forward his tongue should be as the pen of a ready writer, to declare the Praises of the Lord. After this the young man who was sober and composed before, ran up and down, Singing Joy, and calling such Devils as should say anything in opposition.

Harris was not persuaded to convert through reason. It was not the meaning of words that induced him. Instead, the singing and dancing were the cause, physically acting on him. The sound was not seductive, but inductive. His will had no effect on his hearing: He heard whether he wanted to or not. This involuntary aspect of hearing distinguishes it from vision and the tactile sense of motion. He could look up or shut his eyes to not see, he could resist motion by simply standing still, but hear he must. Even stopping up the ears was only a partial solution, as sounds would still have bled in. The sounds of the singing and, to a lesser extent, the dancing were what pried open the door to possession. In one of the Mather's hallmark signs of possession, those sounds went into him and seemingly without volition came back out of his own mouth, causing him to "cry out upon" others and run about "Singing Joy."[22]

Mather used nonverbal vocalization to show the wildness of Case's crew. He treated singing as a cunning display of sonic force. As he described it, the process of conversion denied Harris much agency beyond the initial contact with the "singing Quakers." Mather was warning other Puritans that they should stay away. Neither reason nor grace was enough to protect them if they allowed themselves within earshot of such people. Harris's

new voice posed such a threat that someone pulled his tongue out of his
mouth and killed him shortly after his conversion. For the next few nights,
"the *voice* of this Harris," now disembodied, reportedly awakened Captain
Young, the sheriff in charge of the murder investigation, "loudly demand-
ing justice to be done."[23]

A year after Mary Ross had helped sing Thomas Harris into his final religion, she burned all her clothes and declared out loud that she was Jesus.
Dunham, Ross, and the unnamed married woman then "danced naked together, having nothing but their shirts on," according to Mather. This was
to signify the "the state of the first Adam, in his innocency." Supposedly,
Dunham's wife found out and beat Ross so that she wished not for "Clothes
as a Covering" but for "Armour." While dancing, Ross "uttered such prodigious blasphemy as is not fit to be mentioned." Next, she pretended to be
dead for three days in the Plymouth (now Rhode Island) town of Little
Compton. Upon her "resurrection" she bade the enthralled Dunham to
sacrifice the protesting John Irish's dog, after which they locked him out
of his house, started a fire inside, and shot off a gun. When magistrates demanded that Dunham account for his behavior, he replied that "Mary Ross
bid him, and he had no power to resist." It was this loss of volition to the
voice of Mary Ross that indicated to Mather that these "Quakers" were possessed by the "inmates of hell."[24] Dunham was not persuaded by Mary
Ross's logic or words; the sound of her voice alone compelled him.

Around the same time, according to Mather, an unnamed Plymouth
woman was "howled into their Society, as Harris was." Like the use of "bid"
above, Mather used "howl" transitively in a way that is now obsolete: they
"howled her" much like we now push something. In Mather's world,
sounds were capable of such tangible effects. Other verbs, such as railing
and singing, were similarly transitive in the seventeenth century, whereas
today they are invariably intransitive, their effects now overshadowed and
mediated by the will.[25] They would howl the Plymouth women into their
society; today we would say that they howled *at* her, which then convinced
her to join. Nonverbal aspects of vocalization were not necessarily subject
to reasoned evaluation by hearers during the seventeenth century: they
were expressions of tangible, physical force.

Much like Harris and Dunham, the anonymous Plymouth woman
"quickly fell to railing on others, and then to raving" after she had been
"howled in." That night, however, she and her Quaker company "heard a
very doleful noise, like the crying of a young Child in the yard or field near
the house, which filled the Auditors with some fearful Apprehensions."
The woman involuntarily fled toward the sound, saying "the Lord calls me,
and I must go." The cry—again, a nonverbal vocalization—literally moved

her. Her companions found her shortly afterward, "affrighted and bereaved of understanding." Her husband thought all this was a sign that "the Devil was among them." Mather concurred, concluding largely on the basis of nonlinguistic utterances that "Quakers are under the strong delusions of Satan."[26]

What happened to them? Dunham was whipped and banished from the colony. Mary Ross was less than a model defendant, uttering "uncivell and railing words" at the deputy governor, and then before the whole court she was whipped and sent home to her mother in Boston. The unnamed woman was apparently sent home to her husband.[27]

To Increase Mather, old England was in some ways as near to home as Quaker Pennsylvania. In part, the ties that bound him there were a stock of what David Hall calls "wonder stories," widely circulated tales of God's providence displayed in everyday lives. By relying on them, he was able to seamlessly weave "wild" English Quakers in with his warnings about the Case's crew and Mary Ross, as when he recounted the woes of the spiritually wandering English Puritan Robert Churchman. In 1661, Churchman was "inveigled in Quakerism," according to Mather. "An infernal spirit spake in him, pretending to be an angel of light." It bade him "sing praises; sing praises." Then the disembodied voice commanded him to gather his family, after which, "making use of his [Churchman's] tongue, [it] bid them to lie down and put their mouths in the dust." Later, "the spirit within forced him to sing, [and] sometimes to bark like a dog." He and his family were spared from further molestation as long as Churchman prayed under the direction of a Puritan minister and regularly attended public worship. All went well until he opened a Quaker book to read, after which, among other things, he began having convulsions. During his fits, he supposedly "broke out into these words: 'Thine is the Kingdom! Thine is the Kingdom!' which he repeated (as was judged) above an hundred times. Sometimes he was forced into extream laughter, sometimes into singing." A day later, the spirit finally left him for good, again with the help of a Puritan minister. Mather's entertaining Puritan parables certainly succeeded, and *Remarkable Providences* quickly became a "steady-seller."[28]

While stories of wild Quakers helped bolster a beleaguered sense of Puritan identity in the 1680s, Quakers took exception to Mather's account, and they did so publicly. The vituperative twenty-year-long war of words that followed the publication of Mather's *Illustrious Providences* indicates that more was at stake than meets the eye. The controversy was only partly one of theology. In both Pennsylvania and New England religion and government were intertwined. In any religious government, the prospect of immediate divine revelation poses a threat to social order. Why listen to earthly authorities if one has access to the ultimate authority? In early Protestant theocracies, religious radicals who insisted on unmediated access to knowledge of divine will—like the singing Quakers—posed perceived threats to social order much like anarchists in later centuries.

The problem of placing earthly limits on a heavenly society riddled early Protestant governments because the breakaway religions were in part premised on the rejection of human intermediaries between Christians and their God. Anne Hutchinson claimed her right to speak on this basis. New England Puritans had responded to her unruly voice in the late 1630s with an emphasis on form as a necessity for social order. Puritan secular and religious leaders thought that without the aid of Puritan ministers, people could too easily be fooled into believing that the devil's words were the voice of God. John Winthrop treated Hutchinson's critique of the ministry as a threat to the state. After banishing her and other Antinomians, the brunt of the Puritans' onus shifted to Baptists, and a little later to Quakers. Both groups criticized Puritan formalism, and as a result many were banished, disfranchised, fined, jailed, or even hanged.[29]

Quakers took a different approach to the problem. They also believed in unmediated connection with a higher power. In England this belief had manifested itself in behaviors remarkably similar to those of the "Ranters" they wished would hush. Indeed, the Quaker's founder, George Fox, thought that the ranters almost had it right. Only a few decades before, the English Quakers nearly split along lines of whether or not to take their hats off in prayer, a Ranter hallmark. Not long before that, Quakers (as indeed earlier Puritans) were considered dangerous speakers. They dealt with the problem of heterodoxy and social order by putting one's message of the divine will before the whole meeting for scrutiny, collectively assaying divine will. Individual beliefs always had to be put in the public domain. Consensus and unity were the markers of truly divine communications. In this way, Quakers were able to nominally avoid what Richard Bauman calls the "routinization of charisma" while still tamping down the divisiveness of the ranting spirits.[30] Once Quakers were able to set up their own colony, however, that consensus failed to emerge. They would have to impose it from above if they wanted social order. Consensually hearing God's will in the silence was a delicate matter that was easily disrupted when it concerned government.

Ultimately, the solution for both Puritans and Quakers was to separate church from state. While both denominations eventually did so, they went about it in different ways. In the wake of Salem, the Mathers and other old guard Puritans found themselves somewhat marginalized, but they went down fighting. In contrast, Quakers were divided about how far they should go. After a contentious attempt (discussed below) to formalize

their theology into a politically viable form, the Society of Friends undertook decisions that set them on a long path toward their ultimate with-
drawal from affairs of state. At the outset of the eighteenth century, the
relationship between explicitly denominational governments and de facto
religious diversity had reached a point of inflection. Denominational governments were still a reality of Protestant North America, but from 1700
onward they were increasingly subject to the marketplace of religious
ideas. These insoluble internal contradictions between heterodoxy and or-
thodoxy were the first cracks in the increasingly stressed structures of these
two regional religious governments.[31]

During the 1680s and 1690s, however, Quaker and Puritan theocracies
were battling furiously over their marginal positions as colonial outposts
of British civil society. Both British civil society and its North American off-
spring were changing, but not at the same rates or even in the same direction. In his characterizations of Case's crew, Mather was trying to make
the Quakers appear "wild" in order to place them beyond the pale, in the
process legitimizing the Puritans' own precarious standing in relation to
the British metropole. North American Quakers fought back, however.
The ensuing fracas was as much about political identity as it was about religious belief. They were fighting over who was cultivated and who was wild,
who was Christian and English and who had gone to the devil or the "savages," who was self and who was other.

Both Quakers and Puritans carefully marked who was in and who was
out of their conceptions of civil society using a commonly understood discourse of nonverbal vocalizations.[32] They hedged their positions at the
margins of British civil society, each at the expense of an audibly constructed other. Considering the soundscape makes it possible to locate the
emergence of plural overlapping identities that were distinctly North
American at the end of the seventeenth century. These precarious and
shifting identities were feistily—if not always consciously—deployed in
ways that complicate any monolithic notion of a single "public sphere"
emerging from the confluence of republicanism, print, free markets, ver-
nacularization, and creole identity a century later.[33]

The Quaker George Keith took up the task of rebutting Mather's por-
trait of a debauched and diabolical barking, howling, singing, and dancing Society of Friends. He called Case's crew "plain Ranters." An adviser to
Penn and a well-educated scholar, Keith was not known to mince words.
"Weeping and Howling, and bitter Mourning is more proper" for Case's
crew than singing and dancing, he said, prescribing the appropriate nonverbal vocalizations a people would make upon realizing they had lost their
positions in civil society.[34]

Keith did not deny any of the events Mather described. He never contested the claim that Case's crew barked, howled, or sang innocent Puritans astray. Instead, he argued that the loud troublemakers were not Quakers at all. When authorities publicly whipped Mary Ross and Jonathan Dunham for their escapades at Plymouth, a number of Quakers "openly declared" that they "did not at all *own* them to be of their society." This loud public dis-possession, Keith thought, should have been ample notice that the Ranters were not Friends. Keith found it offensive that Mather did not acknowledge this verbal publication, although he no doubt knew of it. Anyone present at the events, as Mather's "credible" sources ostensibly were, would have heard the "real" Quakers' vocal announcement.[35] The "Tricks and Freaks of Singing and Dancing" that Case's crew and others practiced caused more harm to Quakers than to the would-be converts, argued Keith. Such rowdiness gave Quakers a bad name.

Besides harming their reputations through false association, so-called Ranters had been disrupting Friends' meetings throughout the colonies. Nearly all the Friends who traveled in North America encountered Quakers "that were gone from the Truth, and turn'd Ranters, i.e. Men and Women who would come into Friends Meetings, singing and dancing in a rude Manner." In 1672, George Fox was accosted in Rhode Island. In 1675, William Edmundson confronted "several of those People, that were tainted with the Ranting Spirit" in East Jersey and Long Island. James Dickinson met "Ranters" in Pennsylvania, as did Thomas Story in Connecticut. Nor were Virginia, New Hampshire, or Maryland spared.[36]

Quakers tried to contain the damage done by these Ranters by regulating their utterances in a powerful economy of nonverbal vocalizations. Understanding that economy helps make the threat of Ranterism more comprehensible. Quakers used silence to value sounds more highly. Penn thought it was the faithful Quaker's "Duty to wait upon God in Silence and Patience."[37] Silence in itself was not what was valued. Talk that did not come from an indwelling spiritual experience devalued true speeches. It existed in a sort of mercantilist economy of sounds where utterances were a hoarded specie that was held onto lest it lose value through too much of it circulating too freely, like Spanish silver.

The test of the specie—like biting (or ringing) a coin—was consensus. Speeches that did not come from a place of "truth" wrought division. Penn wrote that God would "restore unto you a pure Language" that would result in unity. Therefore, any guidance that was not in unity with the rest of the Friend's judgments could not be from God.[38] This was a problematic test, however, for truth could not be gauged until the speech had already been uttered.

For this reason, Quaker authors concerned with social order advised that all but leaders ought to remain silent. This embedded a social order within Quaker *practices* that contradicted their ideology of spiritual equality. They were aware of this contradiction between worldly and heavenly means and within a few decades the tensions would become too great for many to bear. The movement would turn inward and away from direct engagement with civil society in the eighteenth century. At the end of the seventeenth century, however, Quakers were running a colonial outpost of British civil society and had to concern themselves with social order.

In *Anarchy of the Ranters,* Robert Barclay consistently marked the quieter boundaries of Quaker civil authority by his attention to nonverbal vocalizations. "Keeping their places," he wrote, the mass of good, silent Christians needed to "shut out the Murmurer." Like Ranters, Barclay's "Murmurers" underscored the relations of powerful sounds to social order: murmurers were sowers of discord so "inwardly vexed" that they could truck with no leader but themselves. "Murmurers at our good Order" were a problem to be rooted out. The threat was that they murmured *at* order, not what they were murmuring *about.*[39]

Murmuring and ranting were closely related, differing mostly in scale and the dangers each presented to social order. Both described nonverbal aspects of vocalization. Murmuring was done in undertones, almost surreptitiously. Its chief danger was that it might distract listeners from true speeches. A second danger was that because it was quiet, it might go undetected, and thus uncorrected. Ranting was fully voiced and louder. It was a full-out sonic attack on the Quaker economy of utterances, one that could not go unnoticed, like murmuring. Keith and other Quakers objected to Ranters—or, as they called themselves, new Quakers—because in turning up the volume and refusing to be silent they posed a direct threat to the underpinnings of Quaker society. In doing so, they transformed a minor social vexation into the threat of anarchy. For this they were banished from civil society.

Quakers banished troublemakers by disowning them. In an aural economy like that of the Quakers, disowning declared the "owners" of disowned utterances to be counterfeit, to be on their own. Disowning restored the value of utterances by excluding the source of the devalued currency without the cruelty of physical punishment, banishment, capital punishment, or imprisonment. Penn maintained that such corporal punishments were the signs of false prophets. Disowning was usually effective. The problem for Keith, Penn, and other Quaker leaders was that the Ranters ignored their social deaths.[40]

Just who was a Ranter was not always clear. Edmundson thought the Massachusetts expatriate Samuel Gorton to be a fine Quaker, but John Bunyeat called him a Ranter. Those labeled Ranters never self-identified as such. Case's crew called themselves "new Quakers." Thomas Story was visiting a Puritan magistrate's home in Connecticut, and the justice's wife asked him about "that Wild and foolish sect aptly called Ranters" because they occasionally preached in the area, always "under the Name of Friends." Keith denounced this widespread practice, saying that "It is not the Name or Profession, that is the sign or mark of distinction" of Quakers. It was instead possession of "the Truth," an inner understanding or light that distinguished them from Ranters. This truth led to "all Sobriety and Gravity in all things, but into none of these mad Gestures, and ungodly Singings and Dancings, under the pretence of Raptures of Heavenly Joy." Perhaps this was so, but Quaker theology had no formal way of silencing Ranters. Nor could they stop them from identifying themselves as Quakers. "And since we cannot oppose them by force," Story gloomily concluded, "they continue to impose upon us in this manner."[41]

Calling Case's crew Ranters worked for Keith, Penn, and the moderate Quakers trying to set up a working English colony by distancing Quaker civil society from its more radical sonic implications. In doing so, however, he may have misrepresented his target as much as Mather's diatribe had. These people did in fact self-identify as Quakers. They never called themselves "*Singing* Quakers," and seldom identified themselves as "Ranters." The reason they were such a threat to Keith and Penn was that they attended and disrupted "regular" Quaker meetings by taking part in them *as Quakers*. In a religion based on "inner light," how could Friends dismiss the words of people claiming to be their fellows? As the Anglican Charles Leslie acerbically pointed out, the "new" Quakers (or Ranters) were doing nothing so much as applying Quaker principles to their fullest.[42] But theology and running a colony were not the same thing, as Penn well knew. Ranterism was not so much about religious belief as it was a commonly recognized name for any ungovernable or anarchic behavior.

The word "rant" evoked a broad web of meanings in the seventeenth century. First and foremost, it was the sound of "heated" speech: foolish, irrational, morally questionable—and, not least of all, dangerous. It could mean a violent scolding, sort of a fit, or, intriguingly, a rim, a margin, or a border, like the half-wild place at the edge of a cultured field. In the North of England and Scotland it could be a rowdy song or a spree of merry-making. In the 1640s the word ranting came to be associated with the public speeches, singing, and deeds of an antinomian sect that arose during the English Civil War. These so-called Ranters were situated on the periphery of English—and later, colonial North American—civil society, with a social analysis that tended to aggravate whomever was writing about them.[43]

The Ranters were loosely organized at best. In the opinion of J. C. Davis, "real" Ranters never existed: they were the invention of frightened polemicists and myth-making historians. Although there may have been no group that self-identified as Ranters, Christopher Hill and others have shown that the term did describe a real, albeit small and chaotic, social movement. By 1653, what movement there was had all but fizzled out.[44] Yet "ranting" was to enjoy a longer life than the movement it originally described, crossing the Atlantic in letters, pamphlets, and manuscripts, and as the cultural baggage of sectarian immigrants.

What exactly did North American Quakers mean when they wrote of Ranters? Thomas Story went to a wedding in Long Island attended by some Ranters. They were for the most part "pretty still," he wrote, "save only an old Man, who sometimes hooted like an Owl and made a ridiculous Noise, as their Manner is." After this nonverbal display, the old "Ranter" spoke out against marriage as an artificially imposed earthly form, and one thus to be avoided. In addition to animal sounds and loud, spontaneous, scolding speeches, singing was a main component of the Ranter's repertoire. Often, however, the songs were without words, rather resembling a spontaneous humming. One practice was to approach potential converts who were alone and surround them while humming a wordless tune. If the potential convert joined in the humming, they would proceed to the next level, possibly preaching in what they thought were tongues but which sounded to outsiders like roaring, a dog's barking, or howling. This might alternate with heated speeches against the adoption of forms, or declamations of joy. Throughout this, the humming and singing would continue and the potential convert would perhaps remain surrounded. The choices were to fight, flee, or join. Joining meant vocalizing—as in Harris's "singing Joy," or it might simply mean making unintelligible noises. Apparently it was not unusual for such converts to switch quickly back to their old allegiances. Perhaps some thought it easier, and perhaps even fun, to join rather than fight.[45]

Women in particular risked being labeled Ranters by the very fact of their speaking publicly. Phyllis Mack maintains that in general, women preachers were one "archetypal symbol of complete disorder" in seventeenth-century England. The Ranters were another. English Ranters, she notes, were also perceived to be the most appreciative of "feminine symbolism and the spiritual power of actual women." A half-century before the emergence of the American Ranters, a complex of fears about the gender and speeches of Anne Hutchinson prompted John Winthrop to write that she "walked by such a rule as cannot stand with the Peace of any State." Even her dying place resonates with Ranterism, for it was from that area that "ranting" seemed to sprout in the late seventeenth century. David

Lovejoy maintains that English and American Ranters were simply antinomians, like Hutchinson. William Penn, echoing Winthrop, defined "The Root of Ranterism" as the assertion that one had no duty "but what thou art persuaded is thy Duty." While the term "Ranter" had not yet been coined when Winthrop wrote, the issues were much the same with Case's crew, including the prominence of women protagonists.[46]

Like the "rant" of half-wild grasses, crops, and weeds that grew only on the edges of cultured fields, Ranters were of but not in civil society. Like those half-wild plants, they threatened to take over the whole cultured lot if left unchecked. But unlike crops, civil society was a contested field in the seventeenth century: where it was depended on whom one asked. As various sects either established themselves more firmly (like the Quakers or Puritans) or faded from the scene, "Ranters" came to stand for anyone who spoke out against the social order *of the accuser*: authors "othered" English and American women and men by calling them Ranters.

In some ways, all of North America was the rant of English society, the half-wild place where it seemed as likely for colonists to lose their Englishness as for them to civilize the wilderness. New Englanders and Pennsylvanians were aware of this, and it raised the stakes in their battle to marginalize each other. In trying to maintain their own memberships in English civil society, they carefully watched for "wildness" from their own margins.

Long Island was the archetypal early American rant. The Dutch and the English fought over official sovereignty for years, finally settling on a border at Oyster Bay in the 1660s, though exactly where was disputed for a few more years. While colonial governments battled over physical boundaries, they more or less left the inhabitants alone. Many malcontents from Puritan Massachusetts and Connecticut drifted in, and left to themselves they became used to independent thinking in affairs of religion and political control. When civil authorities turned their attention to Long Island they found much the same thing as did religious authorities: Long Island was a half-wild place full of sedition and a threat to political order.[47]

At about the same time as Oyster Bay and the rest of Long Island became the hearth of religious Ranterism, civil officials were noting its political unruliness as well. Malcontents from New England, the Jerseys, and Pennsylvania sought and found refuge in the island's towns. The island had been somewhat ungovernable for years, and "for a Long time groaned under many grievous inconveniences, and discouragements occasioned partly from their subjection, partly from their opposiçon to a forraigne Power." This vocable groaning left them in a "distracted condiçon" in which "few or no Lawes could bee putt in due Execuçon, Bounds and Titles to Lands disputed, Civill Libertyes interrupted, and from this Generall Confusion, private dissentions and animosityes, have too much prevailed against Neighborly Love, and Christian Charity." In short, Long Island had succumbed to the same sonic forces that Barclay described in *Anarchy of the Ranters*. The Dutch and the English fought each other and among themselves over visible boundaries, but neither was able to effectively police its vocable margins. After the Dutch capitulated, New York still battled its Long Island inhabitants, requiring them to write a "Draught of each Towne Limits, or such writings as are necessary to evidence the Bounds and Limitts, as well as the right by which they challenge such Bounds and Limits," seeking unsuccessfully to literally draw Long Islanders into the social order.[48]

By the 1680s, Long Islanders' politically seditious behavior had become a threat to English colonial social order in the same way that Ranters threatened religious aspects of civil society. Governor Thomas Dongan recognized that the island had an "abundance of Quakers preachers[,] men[,] and Women especially; Singing Quakers; [and] Ranting Quakers." The terms that the colonial English government in New York City used to evaluate the threat emphasized the sounds of the people in ways that reinforced the descriptions of the threats to religious life. "Ryotous And Tumultuous" assemblies were meeting "without any Lawfull Authority." The governor instructed his New York constables to break up these local governments on the fringe of the colonies in order that "Disquiett may not Happen thereby and the peace and Quiett of his Majties Subjects be Preserved." Reinforcing the vocable goals of limiting Long Islanders' political autonomy, constables were instructed to prosecute the seditious in whatever way brought the "most for the quiett of the Governmt."[49]

During the 1680s, Quakers and Puritans still looked askance at their Western frontiers. The public gaze was predominantly toward the Atlantic, and the public as audience listened for England. Although the eastern littoral ceased to be a frontier in the eighteenth century, Long Island during the mid-to-late seventeenth century was the rant of English civil society in the colonies, not only religiously but politically too. Colonial officials made the connection explicit in their construction of both the normative, "quiet" government soundscape and the "tumultuous" disorder of those who had "groaned" too long under the stress of living at the limits where the "wild" and the "other" met the "civil."

George Keith undertook a campaign of challenging Puritan ministers to publicly recant for calling "Case's Crew" Quakers. He did this in print, in public letters posted Luther-like to meetinghouse doors, and in private letters to the leading Boston ministers. He never got the opportunity to air his differences with Mather in a public aural space, however. He asked the leading Boston ministers for a "publick hearing, or meeting with you, either in one of your publick Meeting Houses or in any other convenient place where all who are desirous to come may have liberty." The Bostonians feared letting the public hear Keith, for as Harris's tragic conversion showed them, Quakers could sing or rail or howl even the soberest Puritans into the Society of Friends. The ministers responded to Keith's request for a public audience by saying that they had "neither list nor leasure to attend his Motions: If he would have a Public Audience let him Print." Puritan leaders felt confident that they could control the visible effects of Quaker objections, but not the audible ones, which were far more powerful. Their strategy apparently worked. As late as 1702, Cotton Mather recounted his father's version of the "singing Quakers" virtually unchanged. Puritans saw no advantage to yielding on any of Keith's points in print. There they could safely ignore his words. Not so in the audible world.[50]

In print, Keith tried to turn Mather's own weapon against him, using the singing Quakers to question Puritan claims for membership in civil society. "The ranting crew of Case's followers" resembled Mather's own Puritans more than the Quakers because they adopted a philosophy much like Calvinist tenets of predestination: the Ranters believed that whatever they did was "foreordained infallibly, and unchangably [by God], whatever cometh to pass." Keith then sullied Mather with a further mark of otherness: Puritans, Keith claimed, followed the practice and example of "the ancient, malicious, and persecuting Heathens," because they made no distinction between nominal and true Christians. Playing what he thought to be his trump card, he said Puritans were not only as wild as the English Ranters, but as wild as "these Rustics, that rose up in war against their lawful Princes in Germany, and the mad crew that followed John of Leiden," referring to the radical Protestant peasant revolts that had so appalled Martin Luther.[51] Thus, he thought he proved, Quakers were the rightful British colonists while Puritans had gone beyond the pale.

While Keith was attempting to best the Puritans in the endgame of membership in British civil society, he also was taking the first steps toward the schisms that would split North American Quakers many times in the eighteenth century. Keith wanted to make the Society of Friends less vulnerable to murmuring and ranting by strengthening its institutional structure, perhaps with himself at the helm. In a colony where church and government were integral to one another, such tactics were more than matters of religion: they constituted the power structures of civil society.

Keith, never the diplomat, quickly mired himself in a tangle of accusations and counter-accusations in trying to protect the Quaker social order from Ranters at one end and Puritans at the other.[52] The controversies over "singing" Quakers left him increasingly convinced that Quaker heterodoxy had to be brought under control. He had William Bradford, who was printing Quaker materials in Philadelphia, publish something of a minimal Quaker creed, with the partial approval of the Rhode Island meeting. The purpose was to provide a guide so that Friends could avoid dangerous speeches. But creeds, especially printed ones, are formal constructions, and some Quakers stood fast in rejecting them. A target of Keith's attacks, William Stockdale, struck back, accusing him of preaching two Christs, one inner and the other an outer form. Keith, incensed at this response, tried to have Stockdale reprimanded by the Yearly Meeting in Philadelphia in 1691. After a long, acrimonious, and inconclusive set of debates involving most of the leaders of Pennsylvania, the Philadelphia Quakers publicly reprimanded Stockdale. They also admonished Keith for his harsh words in dealing with him.

In January 1692, Quaker leaders at Thomas Gardiner's house in Burlington, New Jersey, complained of having to listen to George Keith's "clamour against us, from which he would by no means of persuasion be reclaimed," eating up all of the Yearly Meeting's time. This happened during the time that Quakers were in charge of Pennsylvania (the main yearly meeting took place alternately in Burlington and Philadelphia), so Keith's clamors were a problem for civil as well as spiritual order. The Quakers met the next day, minus Keith, and appointed John Simcock, Griffith Owen, and Samuel Jennings to go and admonish the absent Keith. The meeting subscribed to a letter charging Keith with "filling divers Meetings with his Tedious Clamour and not giving us oppertunity to quietly proceed to any business." Silence was a delicate thing, highly valued at Quaker meetings, and Keith's clamors in a situation where speech was supposed to be kept to a necessary minimum were highly disruptive.[53]

Keith, not content with the outcome, refused to apologize and was thus barred from preaching. Rather than silently acquiescing—as he advised murmurers and Ranters to do—he gathered a significant minority of the Society to his cause and began holding separate meetings. The Society divided into Keithians and Lloydians. Thomas Lloyd was lieutenant governor at the time, the *de facto* leader, while Governor Penn remained in England. The Keithians lashed out at the Lloydians, accusing judges and ministers of drunkenness, gambling, and insolent speech. In 1692, as witchcraft accusations were racking New England's "murmuring generation," the Lloydian meeting disowned Keith and his followers, dividing Quakers one against another.

The Keithians appealed, making twelve points in defense of their actions. Bradford the printer took their side and published the appeal. The Lloydian's promptly arrested him and charged him with unlicensed use of the press. His type and press were confiscated. Another Keithian, John McComb, was charged with distributing the offending pamphlet. The Quaker judges jointly issued a proclamation charging Keith with slanderous speeches, but not against the church: it was sedition against the King and Queen's government, Lieutenant Governor Lloyd, and the magistrates themselves.

Though officially incarcerated, the sheriff let Bradford and McComb free on their own cognizance during their trials. But letters from prison struck a deep chord in Quakers, so when the two prepared a public statement in their defense, they wished to sign it from the jail. The jail adjoined the sheriff's house, and unfortunately for the would-be prison martyrs, the sheriff was away, so they were locked out rather than up. They would have had to break in to authenticate their jailhouse manifesto. Instead of risking another crime, they signed it in the doorway, publishing it along with Keith's defense and the original charges. In the midst of the trials, the Crown replaced Lloyd with their own agent, Deputy Governor Col. Markham. At this point the proceedings ended, with small fines being the only penalties. Bradford's press was returned, and all the "prisoners" were set free.

The Keithian moment had passed, but Keith did not know it. Failing to lead Pennsylvania to an orderly Quaker theocracy, Keith turned to what he imagined to be the center of civil society, England, in order to pursue his cause. But English Quakers were indifferent, some providing lukewarm support for this faraway cause, others ignoring or opposing him. In the meantime, his supporters in North America drifted back to the side of the government, or off to other denominations.

Frustrated by the Quakers, Keith made a last ditch effort to be a player in the center of civil society: He became an Anglican priest. In 1702, he returned to the colonies as a missionary for the society for the Propagation of the Gospel in Foreign Parts. In his mind he must have thought himself the ultimate emissary of the center of civil society, or even civilization itself. In his grasp for the center, however, he succeeded only in completely marginalizing himself. Puritans were amused at their former nemesis in priestly garb. Quakers simply derided him.

In important ways, North America had drifted from the grasp of the metropole by the beginning of the eighteenth century. What emerged in fits and starts, mostly unconsciously, was not an overarching American identity, but a series of English identities that no longer fit the political situation in England. Long before a single imagined American community emerged, disjoint aural communities created places in which public opinion could be formed and heard. Nonverbal aspects of the vocal soundscape operated in a commonly understood framework, with loud nonverbal vocalizations marking membership, possession, otherness, and wildness. A "public hearing" rather than a "public sphere" was the most important resource that theocratic leaders sought to protect (or gain access to, as the case might be). Thus, if the much-vaunted public sphere posited by Warner and others emerged later in the century as what Anderson calls an "imagined community," it had to do so not *ab nihilo,* or from a British identity, per se, but from many American identities, each of which constructed its own version of civil society. These identities were not self-consciously fashioned as American. Rather, they were failed attempts at being British when they had diverged too much to ever recover that. England was too far away to do much about it, and the North American colonies were in many ways too marginal for them to care much. Perhaps plural American identities need to be given a hearing on their own terms as an alternative to a single national identity arising later in the eighteenth century Soundways allow us to leave America plural, too, because various identities could be in tension with each other without necessarily resolving into one, like a composer's use of harmony and dissonance.

By the beginning of the eighteenth century, the threat of ranting had all but disappeared. In 1705, the Anglican missionary John Thomas was sent to Long Island. He reported to his Society for the Propogation of the Gospel supervisors in London that the people of Oyster Bay "have generally been canting Quakers, but now their society is much broke and scattered."[54] Sarah Kemble Knight, during a journey through Connecticut in 1704, recounted the story of a group of singing Quakers who came to visit one of their fellows in those parts. The man was not home, but his wife, who was "not at all affected that way," sprang into action when they "sat down (to the woman's no small vexation) Humming and singing and groneing after their conjuring way. Says the woman are you singing quakers? Yea says They—Then take my squalling Brat of a child here and sing to it says she for I have split my throat with singing to him and can't get the Rogue to sleep."[55] In Kemble's humorous tale, this sent them packing.

Keith fared no better. To defend themselves the Friends simply quoted his old and able defenses of Quakerism back at him, even though he wrote pamphlets taking back everything that he had formerly said. Once in print, he found he could not "unsay" things easily.[56] Quakers roundly ignored his objections and retractions. One bit of Quaker doggerel irked the Anglican Keith so much that he made the mistake of reprinting it in order to refute it, thus giving it wider circulation than it ever would have had otherwise:

> But the Light hath still Triumphed ever all
> For Ages past and Triumph ever shall:
> Whilst baffled Keith, who better Things once knew,
> May Rave ith' Dark with his Benighted Crew.[57]

Keith—even with a "crew" (echoing Case's, no doubt)—was no longer a threat. His voice was not even a murmur, much less a rant. It was the harmless raving of a powerless and mad old man, completely outside the workings of civil society. Nothing could have hurt him more.

And if we listen to the soundscapes that New Englanders and Pennsylvanians constructed at the end of the seventeenth century, we will hear contentious, plural, squabbling civil societies that had not yet been drawn into a public sphere. Though not yet American, they were vocably something other than the English men and women they were trying so desperately to remain.

The Howling Wilderness

Seventeen years later I was in another band, this one somewhat mellower. Two of our members were Mohawks, and we would play at various cultural festivals. At one, in Akwesasne Territory on the Saint Lawrence Seaway, the outdoor portion of the concert was shut down by a thunderstorm. The number of people dwindled, mostly to those in the other bands or local people. One of the headline bands, the women's a capella group Ulali, decided to perform inside the community center on the site. They chose a location in one of the hallways with a high, vaulted ceiling. Their voices, accompanied only by a heartbeat drum, filled the hall, sending shivers down my spine, and I am sure down others' as well, even though the songs were often in languages many of us did not know. "Ulali" is the Tuscarora word for the wood thrush, a magnificent songbird, and the name of a Tuscarora woman who had been renowned for her singing. Ulali described themselves as three First Nations women who "sing music in the many styles and languages of our ancestors in the western hemisphere. We do not call ourselves 'Native American' because our blood and people were here long before this land was called the Americas. We are older than America can ever be and do not know the borders. Our brothers and sisters run from North to South and into and under the waters for miles and years back."[1]

I n the previous chapter, we explored some of the ways English American settlers included and marginalized each other from civil society. This chapter expands upon the notion of "othering" to listen to the vocable and paralinguistic soundscapes of people usually placed beyond the rant of civil society, Native Americans. We will briefly consider the sonic strategies that white invaders used to exclude First Nations people from the societies they were building on native ground. Then we turn to how Native Americans used vocable and paralinguistic sounds to structure their own societies and their relations to the colonists.

Unlike the previous chapters, this chapter spans the seventeenth and eighteenth centuries. There are two reasons why. First, there are marked continuities in the ways First Nations people expressed and interpreted voice across the centuries. The changes that did take place were more in the realm of how white colonists treated cross-cultural vocal exchanges rather than how Native Americans treated them. Colonists became less concerned with Indian sonic protocols as the balance of North American power relations shifted in their favor. Indian practices remained fairly consistent. Second, the source materials for both centuries taken together create a richer and more nuanced history. Looking at changes in practices over time helps to expose biases as the balance of power shifted. Soundways that whites held less valuable were dispensed with over time, indicating Native American ways of approaching cross-cultural situations that the colonists no longer needed to attend to so carefully, if at all.

A Wilderness That Howled

In his famous essay, "Errand into the Wilderness," Perry Miller makes a reduction, focusing on the Puritans' "errand" as a metaphor for their mission in New England. He treats the wilderness as a simple blank, the more or less ambient space into which the Puritans pursued that errand. Miller sought to correct Frederick Jackson Turner's frontier hypothesis, in which American exceptionalism sprang forth from the availability of "free" land, and which also assumes a wilderness waiting to be populated. But regardless of whether it was errand or opportunity, the Puritan wilderness was not empty.[2]

The seventeenth-century wilderness did not exist within the limits of European-style cultivation and culture. It was a landscape, however, and thus the human construction of those who lived both in it and near it. Early Americans neither held nor had available the modern concept of a neutral, uninhabited environment as "nature." For First Nations people, humans held the same valence as the rest of nature. For the colonists, wilderness was a part of nature broadly defined; it was a place where distinctions between humanity and environment, good and evil, or high and low were elided indiscriminately. It was a place where one's proper belonging—to a community or God—did not hold, a place where "will" (possibly the root of "wild") overcame belonging. It was inhabited by wild "others." Those not belonging to a godly community belonged by default to the devil. To the colonists, the vast unimproved forests of eastern North America were the epitome of wild, as were its inhabitants, whether human, animal, or spirit. Transgressing boundaries of civil society and individual minds, the wilderness was the constant threat that wildness, willfulness, and disorder would colonize fragile English ways instead of vice versa. To go into the wilderness was risky, as was trying to carve out a community in it. It was "bewildering" to those lost in it, and it was a difficult place from which to return.[3]

Puritan chroniclers used the howling or roaring wilderness to hold down the outer reaches of their landscapes. In Deuteronomy, the Hebrew God is said to have plucked Jacob out of such a "waste howling wilderness" to found Israel, so the Puritan errand into the wilderness was a symbolic return to the place where God had chosen his people, perhaps in hope of an encore. Miller contends that the first generation's errand was successful, but that like Jacob's heirs, the second generation faltered, bewildered. In the 1650s, Thomas Hooker warned his second-generation congregation that they would have to "come into and go through a vast and roaring wilderness." In 1654, Edward Johnson would write of the founding of Concord, "Thus this poore people populate this howling Desart." So too, when the Puritans banished Roger Williams to what he called the "howling wilderness," the biblical reference would not have been lost on many Puritans. Williams lived there not alone, but with Native Americans.[4]

What was a wilderness that it howled? In 1662, the Puritan poet Michael Wigglesworth wrote of the "waste and howling wilderness" ever threatening to overrun Puritan life as a place where "Where none inhabited / But hellish fiends, and brutish men / That devils worshiped."[5] What howled most often in colonists' perceptions of their wilderness were not winds or wolves, but Indians. Colonists did not have to imagine the voices of those they fashioned their wilderness demons.

Mary Rowlandson, wife of a prominent Puritan minister, described her trek "into a vast and howling wilderness" as a captive of the Narragansetts during Metacom's War. She wrote in part to distance herself from her captors, to show that she herself had not gone wild, signaling her restoration to civil society by publishing her narrative. The landscape she described was by no means empty. "The Indians were as thick as the trees," she observed, explicitly comparing the wilderness of the forest with the human wilderness in which she perceived herself.[6]

For Rowlandson, what howled in the wilderness was its people. She recalled how "a company of hell-hounds" came out of the forest "roaring, singing, ranting, and insulting" as they killed and captured the inhabitants of Lancaster, a small Massachusetts frontier outpost. Her first night as a captive was "the dolefullest night that ever my eyes saw. Oh the roaring, and singing and dancing, and yelling of those black creatures in the night, which made the place a lively resemblance of hell." And hell, in the imagination of Increase Mather, the most illustrious of Rowlandson's cohort of

Puritans, was best captured by "the howlings and the Torments of the pitt beneath." Later, another band of Narragansetts met her party with an "outrageous roaring and hooping," thus signaling from as far as a mile away how many English they had killed, giving "a shout so that the very Earth rung again." Again, Rowlandson related their frightening power to the nonverbal (at least to her) sounds they made. When her (female) master's infant died, Narragansett mourners came to "howl with her." This phrase notified Puritans that the women were invoking hell as they marked the infant's passage to the invisible world. Howling, roaring, yelling, and whooping were all markers of wildness to Rowlandson, demonstrating both the incivility of the Indians as well as—by contrast—her own unbroken cultured demeanor in the face of these putative challenges.[7] Demonstrating her own civility was crucial to regaining her position in Puritan society. She achieved that in her narrative by marking the otherness of, and then distancing herself from, Indian ways.

Such descriptions were not limited to the Puritans. While a captive of the Powhatan Indians, John Smith described their religious ceremonies as "howling devotions" made by "devils." At another point he described their singing as "howling" led by a "great grim fellow" masked with skins and feathers who led with a "hellish voyce and a rattle in his hand." Strachey corroborated Smith's estimation of the men, calling a similar performance he witnessed as little more than "showting, howling and stamping their feet." In contrast, Strachey thought the women had a "delightful and pleasant tang in their voyces" when they sang. Smith differed in his opinion, describing the same singing as an "excellent ill varietie" of the "most hellish cries, and shouts," every bit as diabolical and terrifying as that of the men.[8] Smith and Strachey were much more dependent on this wilderness than the second generation Puritans were on theirs. Nonetheless, the Jamestown men marked the Powhatans' "otherness" by rendering prayer, singing, and language as vocable "howling."

In the days before telephones and radios, long distance sonic communication was limited by loudness. A loud signaling vocal sound was called a "halloo." This is the ancestor of today's ubiquitous "hello," a word that has only become a greeting since the rise of the telephone. In early America it was used as a locating device in the woods or the wilderness. Although from different roots, "halloing" was often used interchangeably with "howl" in reference to the sounds First Nations people made. So right before John Smith was captured by Opecanchanough, he heard "a loud cry and a halloing of Indians." Using the term in its more traditional sense, Strachey noted that Powhatan had about forty or fifty sentinels who "at every half howre . . . doth hollow, unto whome every Sentinell returns answere." Over a century later, when the African American Briton Hammon was captured by Florida Indians, he reported that they made a "prodigious shouting and hallowing like so many Devils."[9] Here "hallowing" served the same function as "howling," namely, marking Indians as other, and by inference, Hammon as a rightful member of civil society.

By the eighteenth century, the wilderness in some respects had become whatever was beyond the pale of one's own social imagination. For example, the Scottish-born gentleman Alexander Hamilton, traveling by ship along the Hudson River, met one Hugh Wilson, "an impudent fellow" who came on board from a canoe and accosted him with "How do you do, country-man." Hamilton discovered him to be a "genuine Teague" by "his howl in singing the Black Jock to the negro fiddle." "Teague" was a term of derision for a Roman Catholic Irishman, particularly favored as a sobriquet by the Scottish colonists of Northern Ireland. So in one way, Hamilton was claiming membership for himself to Scotland and thus to British civil society while at the same time placing Wilson in the "wilds" of colonial Ireland. This reading is complicated by the fact that they were in the "wilds" of colonial New York as well. There, Wilson had attempted to cross not only distant territorial lines, but also, and more immanently, class lines. The ethnic crossing was a source of amusement or irritation. In contrast, the transgression of class lines was more of a threat in the wilderness, where "civil" distinctions like social standing were vulnerable to just this sort of leveling. Smarting from the attack on his class status, Hamilton attributed a howl to Wilson that placed him beyond the bounds of civil discourse. He furthered this othering by bringing in the sounds of race. Hamilton located the Irishman as the more natural colleague of the "negro fellow on board who told me he was a piece of a fiddler and played some scraping tunes on one." A violin in the hands of a gentleman became a fiddle in the hands of everyone else, and a "negro fiddle" in the hands of the African American passenger. By placing Wilson's Irish "howl" against the backdrop of an African American—and thus probably syncretically Africanized—version of a Scottish jig, Hamilton fended off the perceived transgressions on civility by locating himself at the center of the project of empire, colonial class relations, and the American racial project, while sonically relegating Wilson—and the African American—to the proximate wilderness of those same projects.[10]

By the nineteenth century the howling wilderness had fallen silent. It was populated only by animals and imagination. Another domain of powerful sounds had been reduced to metaphor, so much so that Henry David

Thoreau could report from the Maine woods with his Indian guide in 1857 that "generally speaking, a howling wilderness does not howl: it is the imagination of the traveler that does the howling."[11]

Native American Vocable Sounds

The Puritan missionary John Eliot, in setting out his grammar of the Massachusetts language, created a typology of vocal sounds. He did so comparatively, by setting off Massachusetts from English; and he did so by generalizing about paralinguistic and extralinguistic sounds. The Massachusetts Indians were always "they," the sonic other, juxtaposed with "we," Eliot's would-be missionaries. Eliot also used linguistic sounds, separated out from their meanings, as part of his civilizing project. He subsumed the sounds of Indian languages under European languages, saying the latter's alphabet was sufficient to the task of describing the sounds of the former.[12] The Massachusetts language was thus wholly contained within the dimensions of English. This act of containment was a prerequisite for colonization, in this case linguistic colonization. Just as Puritans thought Indian bodies had to be put within English-style "Indian towns," so the sounds of language had to be placed inside the sounds of the English language. That which fell outside was wild, dangerous, and uncivil. It was howling and ululation.

A major component of Eliot's missionary (and colonizing) activity was to translate the Book of Psalms into Massachusetts, set it to meter, and have the Indians sing it "in our Musicall Tone" rather than in their uncivil "yelling." Besides linguistic sounds, he relates, "We" have the sounds of "L'tation [lamentation] and Joy: of which kinde of sounds our Musick and Song is made" while "they" have the sounds of "Ululation, Howling, Yelling, or Mourning: and of that kinde of sound is their Musick and Song made." He elaborates by saying that "They have Harmony and Tunes which they sing, but they matter it not in Meter." Their musical voice, though tuneful, was made from the "kind of sound [in which] they also hallow and call when they are most vociferous." Their singing was in the same tone as "a word or a sound that uttereth the passion of the minde, without dependance on other words." "Our" music was rational, contained by language, and subject to rules and meter, while "theirs" was emotional, outside of language, difficult to reduce to rules (Eliot was never able to complete his grammar), and without meter.[13]

Indians and Europeans took careful stock of how each others' speech sounded. In part this may have been a function of incomprehension. In a world where one knew speech was taking place but did not understand the language, the sense of hearing would have been in a state of heightened arousal, as the mind searched for the meanings that the hearers knew were there but could not comprehend. The paralinguistic elements of speech would stand out in particular, being processed by a different part of the brain and often assumed to be more or less universally understandable.

For example, the Jamestown colonists carefully noticed the timbre of Native American speech. Smith described the paralanguage of the Iroquoian-speaking "Sasquesahanocks" as otherworldly, "sounding from them, as it were a great voice in a vault, or cave, as an Eccho." In contrast, he described their neighbors, the Potomacs, as "showting, yelling, and crying, as we rather supposed them to be so many divels." George Percy thought that Powhatan's people made "a noise like so many Wolves or Devils" in their speech. He described their formal speeches as a "foule noise" and said that in worshiping the sun they made a "Hellish noise foaming at the mouth."[14] In marking these differences, Smith and Percy placed Indian languages in the realm of the irrational, the willful, the wild.

Eliot interpreted Massachusetts Indian personality traits from the vowel sounds of their language:

> It seems their desires are slow but strong;
> Because they be utter'd double-breath'd, and long.

The Puritan minister Henry Whitfeld, who understood nothing of the language he was hearing, nonetheless felt he could ably judge an Indian's public prayer by his "voyce and outward deportment." Eliot noted that for the Massachusetts, it was considered great art to compound one's words into long words. The Indians supposedly did this mostly because they liked the sound, which was also why they occasionally added extra syllables not required by the grammar.[15]

If for a moment we step further afield, to the Mexica (Aztecs), we can perhaps glean a glimpse of how First Nations people themselves thought they should sound. Mexica ideals were gendered. Girls were to speak unhurriedly and plainly, in an even tone, without "squeaking" or "murmuring." Boys were "to speak very slowly, very deliberately," taking care "not to pant, nor to squeak, lest it be said of thee that thou art a groaner, a growler, a squeaker." Crying out marked one "as an imbecile, a shameless one, a rustic, very much a rustic."[16] As we shall see, these prescriptions and proscriptions, particularly the last, parallel how Indians further north thought they should sound.

Refreshingly, the Recollect missionary Père Chrestien LeClerq recorded what Indians thought of the Europeans. Micmac Indians thought that the sounds of French speech fell far from their ideal. LeClerq recounted how

> Indians never interrupt the one who is speaking, and they condemn, with reason, those dialogues and those indiscreet and irregular conversations where each one of the company wishes to give his ideas without having the patience to listen to those of the others. It is, accordingly, for this reason that they compare us to ducks and geese, which cry out, say they, and which talk all together like the French.[17]

All of these descriptions were comparative. Europeans compared in order to judge how human the Indians were, whether or not it was possible to convert them, and what they were essentially like. Indians used paralinguistic aspects of speech much the same way the Europeans did, to properly situate the speakers in relation to society. They too used paralinguistics to judge (negatively in this case) the humanity of the Europeans, comparing them to wildfowl.

While Eliot's and others' descriptions of First Nations vocable sounds in these contexts tell us more at first glance about ideologies of colonization and conversion than about Indian soundways, the latter can be pulled out of the documents with some confidence. Battles, treaties, and other negotiations with Indians played important roles in colonial statecraft. First Nations people were critical forces, and the care that colonists put into their relations with them reflects this. Treaties, ethnographies, captivity narratives, and other documents expose a coherent set of practices that reappear too consistently to be fabricated.

Piskaret's Revenge

First Nations people throughout the eastern woodlands shaped and governed their battles through extralinguistic and paralinguistic sounds. Chance encounters between hostile parties had sonic protocols. Planned battles were prepared for with war songs. Attacks were initiated with yells. Victories and losses, both during and after battles, were marked out loud. The sounds they made in these circumstances were part of the very identity of aggression. A tale of revenge and international intrigue illustrates many of these practices.

During the late sixteenth century, the Algonquins were a fierce, warlike nation of hunters who traded with—and looked down upon—their farming neighbors to the south, the Five Nations or Iroquois. During a time of scarcity, the Algonquins took several parties of Iroquois hunting with them to perform the drudge work. The Iroquois went in the hopes of better learning to hunt. One party of six Algonquins and six Iroquois were obliged to split up by the scarcity and cold. The Algonquins assumed the Iroquois would starve to death, but instead—due to their superior patience and fortitude—they returned having killed several "wild cows" while the Algonquins remained empty-handed. Jealous even though the Iroquois shared all and gave them the best pieces, the Algonquins murdered the newly successful hunters and buried them, claiming that they split up in the woods and the Iroquois never returned. The Iroquois sent out a party to investigate, retracing the hunters' tracks, and discovered the corpses, which had been dug up by wild animals. From their remains it was obvious they had been murdered, which greatly upset the Iroquois and instilled a spirit of revenge in them. Nonetheless they were weaker than the Algonquins in war, so they ended their alliance and retreated southward.[18]

Into this situation the French arrived, and under Samuel de Champlain they somewhat arbitrarily allied themselves with the Algonquins, forever earning the enmity of parts of the Five Nations. Champlain, in order to show French valor as well as his commitment to the new alliance, led a party of Algonquins across the lake that now bears his name. "They had not long been in the Lake," writes Colden, "before they discover'd a Body of the Five Nations going to War. As soon as they saw each other, Shouts and Crys began on both Sides."[19]

Shouts and cries were the standard way of beginning a battle, whether on chance encounter or in a planned attack. Throughout the eastern half of North America, First Nations people began their battles with a loud yell. In the Chesapeake, Jamestown settlers knew that the Powhatans and their neighbors cried out "their accustomed tune which they use only in warres." This sounded to John Smith like "horrible shouts and screeches, as though so many infernal helhounds could not have made them more horrible." Smith tried to outdo the Indians by expressing the awful din of European wars to Powhatan himself, noting that the Indian king was duly impressed. In September 1677, Quintin Stockwell described the onset of an Indian attack on the New England frontier town of Deerfield. He and another man "ran away at the outcry that the Indians made," thus saving themselves. Swerisse, an Oneida sachem, noted in 1689 that war shouts "vehemently raised" the spirits of his warriors, readying them to attack on a moment's notice. Also in 1689, the young Puritan John Gyles learned of an Indian attack by the "Yelling of the Indians, the Whistling of their Shot." Footnoting the description several decades later, Gyles—or his editor—remarked that the Indians began their attacks with "a custom of uttering the most horrid howl when they discharge guns, designing thereby to terrify those whom they fight against."[20] Thus the yelling by both the Algonquins and the Iroquois that Champlain heard was part of a widely recognized way of beginning battles.

Although both the Iroquois and the Algonquins gave the war cry, it was too close to dark to begin a battle, so the two sides agreed to put off the fighting until daylight. "The night passed," recounted Cadwallader Colden, "in Dancing and War Songs, mixed with a thousand Reproaches against each other." The reproaches were in mutually unintelligible languages. Each side knew what the other was saying not because of content, but because Indian war songs all over were distinctively recognizable by their sound. For example, Chesapeake warriors would prepare themselves for battle with loud percussive music and singing:

> For their warres they have a great deepe platter of wood. They cover the mouth thereof with a skin, at each corner they tie a walnut, which meeting on the backside neere the bottome, with a small rope they twitch them togither till it be so tought and stiffe, that they may beat upon it as upon a drumme. But their chief instruments are Rattels made of small gourds or Pumpion shels. Of these they have Basse, Tenor, Countertenor, Meane, and Trible. These mingled with their voices sometimes 20 or 30 togither, make such a terrible noise as would affright then delight any man.

The songs could be about their foes, as in the "kynd of angry song" that was sung about the colonists in "homely rymes" that concluded with "a kynd of Petition unto their Okeus [malevolent deities], and to all the host of their Idolls, to plague the Tassantasses (for so they call us) and their posterityes." Pequot Indians were said to have spent the night before battling a combined force of Puritans and their Indian allies "singing and dancing until midnight." In this case, the war songs were to no avail, as the Puritans and their allies circled the Pequot fort and set it afire. Six hundred Pequot men, women, and children were killed, either in the conflagration, or by getting shot as they fled.[21]

War songs were not solely about invoking protection and riling up the warriors. The songs also acted as histories. According to William Smith, writing in the mid-eighteenth century, the Iroquois would assemble on the night before a battle and "having their Faces painted in the most frightful Manner, as they always have in the Day of Battle, every Warrior, rising up in his Turn, sings the great Exploits of his Ancestors, together with his own." The war song was accompanied by a war dance. The combination of the two served to "represent the manner in which those great Actions were performed, which are the Subject of their Song." All the others present formed a "Chorus" that applauds "every notable Act." Smith went on to note that "they represent their great Actions on Trees, near their great Castles, which, together with these War-songs, transmits the history of their illustrious achievements from Generation to Generation."[22]

War songs bridged cultural divides and sealed alliances, too. Often envoys to another Indian nation would learn that nation's songs and bring them back. Even though the language would be different, the song would act as a piece of the alliance. Almost by definition the songs were meant to be cross-cultural, to be understood by enemies or allies who spoke other languages. Gregory Evans Dowd shows how, in 1776, Shawnees, Ottawas, and Mohawks traveled to the southern Appalachians to convince the Cherokees to join them in war. Dragging Canoe, the Cherokees' leader, accepted the northern belt of Wampum and the speech it came with. He was willing to fight, though the Cherokees as a whole were divided into neutralists and fighters in regard to whites. Toward the end of the meeting, a Cherokee leader who had lived among the Mohawks for some time took the belt from Dragging Canoe and sang a war song, perhaps a Mohawk one. The Northerners joined in on the chorus. The alliance and the song went beyond tribalism. It was, according to Dowd, "the amplification of native calls for a pan-Indian alliance." "The striking fact," comments Dowd, "is that Native Americans themselves, unlike many of their historians, could think continentally."[23] One of the ways they did so was through trading the sounds of their respective war songs.

The Algonquins probably thought it strange that Champlain and the French did not participate in their war songs the night before they fought the Iroquois. Nonetheless, battle lines were drawn the next morning, and three Iroquois captains led their troops toward the Algonquins. Instead of firing arrows at the armored captains, the Algonquins stepped aside to reveal the French and their firearms. This surprised the Iroquois, and the French mowed down the three captains, their guns piercing the arrow-proof armor they wore. Then the Algonquins gave the standard "terrible Shout" and attacked. A second volley from the French turned it into a rout.[24]

The Algonquins, their confidence bolstered by the firearms of their new allies, declared open war against all the Iroquois, becoming rash and insolent in their attacks. The Iroquois responded defensively, and slowly wasted away the Algonquin forces through attrition and ambushes. One of these ambushes killed the leader of the Algonquin warriors. By this time the tide had turned and they could only have hopes of revenge, no longer of conquest.[25]

One of their leaders, Piskaret, set out on one such revenge mission with four other warriors and fifteen French muskets in a single canoe. Beforehand, he had wired together the guns in groups of three and gave one set, each with two bullets in it, to each warrior. A party of fifty Iroquois in five canoes came upon them. At first the Iroquois fled, but upon finding that no more Algonquins followed, they came back and "as soon as they came within call, they raised their War-Shout, which they call Sassakue, and bid Piskaret and his Fellows Surrender." Piskaret pretended to give in, saying "he could no longer survive the Captain they had burnt." He then asked only to die in battle so as not to appear cowardly and called for the five canoes to come out to the middle of the river. Once in the middle of the river, Piskaret feinted an escape, causing the Iroquois to separate from one another and surround his canoe. The Algonquins, "the better to amuse the Enemy, sung their Death Song, as ready to surrender themselves."[26]

Piskaret's death song was part of a set of sonic practices around death and captivity. These in turn were part of a broader system of First Nations soundways in which sounds, whether natural, instrumental, or vocal, were integral to identity. Captives of eastern woodlands war parties, as was Piskaret, were expected to sing their identities into being. Native American captors and captives alike treated this practice seriously. It was meant as an expression of identity that captors sought to break via torture, smoothing the way for possible adoption or else a grizzly execution meant to either replace or vindicate the loss of a dead warrior or other family member.[27] Like war songs, the soundways of torture practices and death songs were remarkably similar among many Native American nations because warfare and captive-taking was an important form of cross-cultural contact and sharing, albeit an unpleasant one.

For the captive to keep singing the song—without crying out at inflicted torture—was a way of continuing to wage war by maintaining his or her past identity. Iroquois "have been known to continue singing their Exploits, and triumphing in their glorious Fate, even in the midst of Torments and the agonizing Throws of Death."[28] "Unbroken" behavior would be to respond to the torture with one's own song, the voice never cracking or modulating in response to the various pains inflicted by the captors. Suffering this way was respected, and occasionally it might result in adoption into one of the captor's families rather than death. The sounds one made or did not make under duress were core markers of identity for First Nations people, whether the captive survived them or not.

Carver, recounting his brushes with Algonquin and Iroquois warriors, noted that those Indians "oblige their prisoners to sing their death-song, which generally consists of these or similar sentences: I am going to die, I am about to suffer; but I will bear the severest tortures my enemies can inflict with becoming fortitude. I will die like a brave man, and shall then go to join the chiefs that have suffered on the same account."[29]

The French eventually took part in these tortures too, as a French and Indian force tortured and killed a number of Iroquois warriors in 1684. As the Iroquois were tortured, "they continued singing in their Country manner, and up-braiding the French with their Perfidy and Ungratitude." In 1690, the French sentenced two Indian captives to die at the hands of the French Indians. The prisoners "began to prepare for Death in their own Country Manner, by singing their Death Song." One managed to kill himself. The other refused, and, recounts Colden,

> While they were torturing him, he continued singing, that he was a Warrior brave and without Fear; that the most cruel Death could not shake his Courage; that the most cruel Torment should not draw an indecent Expression from him; that his Comrade was a Coward, a Scandal to the Five Nations, who had killed himself for fear of Pain; that he had the Comfort to reflect, that he had made many Frenchmen suffer as he did now. He fully verified his Words, for the most violent Torment could not force the least Complaint from him, though his Executioners tried their utmost Skill to do it.[30]

One captured warrior was at first adopted, and once he thought himself safe his new family changed their mind and had him tortured. The suddenness of the decision and the surprise caught him off guard, and he failed to muster the courage it took to face his tormentors with a death song. "It was fearful to hear him shrieking in the dead of night. He shed great tears, contrary to the usual custom, the victim commonly glorying to be burned limb by limb, and opening his lips only to sing."[31]

English-speaking captives were also expected to perform a song, and if they could not master the pronunciations taught by their captors, they could sing in English. The words were not the assertion of identity; the sounds were. It is interesting that captors held this to be an important practice, though obviously European captives did not. While most singing appears to have had a gendered pattern in northern First Nations societies, women as well as men were expected to sing when captured.[32]

The Jamestown colonists were abject failures in this department. They "broke" immediately. The Powhatan Indians made a "scornefull song" about them, the refrain of which went, "*Whe, whe, yah, ha, ha, ne, he, wittowa, wittowa.*" It derided the colonists for immediately crying out "whe, whe" upon being tortured or even seeing torture, which, according to Strachey, "they mock't us for and cryed agayne to us Yah, ha ha, Tewittaw, Tewittawa, Tewittawa: for it is true they never bemoane themselves, nor cry out, giving up so much as a groane for any death how cruell soever and full of Torment." New Englanders fared no better. After Pequots captured and killed four men in 1636, they surrounded the fort where the survivors had fled and "challenged them to fight, mocking them in the groans and pious invocations of their friends they had just tortured." New Englander Joseph Tilly was captured and tortured to death, "but as nothing which they inflicted upon him excited a groan, they pronounced him a stout man."[33] Here there is no mention of song. It may or may not have been part of the proceedings. The concept remains the same, though. To break a captive would be to elicit involuntary, nonverbal vocalizations. Since Tilly remained in control of his voice, he was unbroken.

So in singing his death song, Piskaret and his would-be Iroquois tormentors were engaging in a well-understood test of his and his companions' identities. But it was a sham for their amusement, says Colden, for then, by plan, each of the five Algonquin warriors raised his guns and fired "between Wind and Water" to shred the Iroquois' birchbark canoes. The five Algonquins then batted in the heads of all but the leaders as they swam from their sunken canoes. The leaders were taken prisoner and burned alive, probably singing their death songs instead of causing them to be sung.[34]

Piskaret was not yet satisfied, so after returning with the warriors, he set out alone toward an Iroquois town. At night he went from house to house and silently tomahawked as many Iroquois as he could. Then instead of running away, he hid in a woodpile at the edge of the village and set out that night to kill more. He was less successful the second night, but still managed to kill two of the warriors on guard and sneak off into the night. The next day he called out to the village and the Iroquois sent out a party to catch him. He easily outran them and would then double back and call them to confuse them. After a day of this, the warriors were exhausted and set up camp and fell asleep, upon which Piskaret killed them all. That apparently glutted his revenge.[35]

A few months later a large Iroquois army set out for Algonquin territory. A small scouting party went ahead and encountered Piskaret. "As they came near him," Colden reported, "they sang their Song of Peace, and Piskaret taking them for Ambassadors, stopt, and sung his."[36] This was enough for Piskaret to think he was safe and he invited the scouts to his home, telling them of the whereabouts of the rest of the Algonquins along the way. They ambushed and killed Piskaret and went on to rout the Algonquin army based on his intelligence. Although this instance of the peace songs failed, the fact that Piskaret was trusting of the song's efficacy indicates its normative status as an international relations tool. The songs were in mutually unintelligible languages—Algonquian and Iroquoian being two distinct linguistic families—so the content of the songs was not at stake. The genre, tune, and timbre—in short, their sounds—are what distinguished them as peace songs and effectively served the Iroquois in setting up their ambush.

Peace songs remained a vital part of Indian diplomacy throughout the colonial period, in relations with whites as well as with other First Nations people. In Albany, at the end of treaty negotiations with the English in 1684, "the Oneydoes, Onnondagas and Cayugas, joyntly, sang the Peace Song, with Demonstrations of much Joy; and Thank'd the Governor of New-York for his effectual Mediation with the Governor of Virginia, in their favour." In the mid-eighteenth century, William Smith noted that in times of peace, Iroquois civility took a musical turn: "their young Men and Maids, of fine natural Parts, with their Bards, or Druids, should frame Songs of Peace, when their venerable Sachems mention it, that their Speech seems a poetical Language, or rather a Sort of Divine Enthusiasm."[37] "Enthusiasm" in its Enlightenment context had a somewhat negative connotation associated with the shaking of Quakers and the ecstasies of Great Awakening evangelicals, so whereas the Quakers were relegated to the wilds along with the Indians in the seventeenth century, the Indians, no longer a part of the wild, were relegated to the same status as other "irrational" enthusiasts in the eighteenth.

Antiphony and Singing

Sounds sealed treaties and made them real, whether between Indian Nations or between Indians and whites. First Nations people structured their public spaces through nonverbal response to speech. The chief public space was the council, which served as the center for treaty-making and war-making. Councils could be held within a nation or between nations, whether Native American or European. They were gendered, age-differentiated spaces, with women and children generally excluded, and elders speaking first, perhaps followed by some of the younger warriors. The speeches were assented to with a loud communal vocalization.

The expression of assent, which, according to Jonathan Carver, "they repeat at the end of almost every period, is by uttering a kind of forcible aspiration, which sounds like an union of the letter OAH." Eighteenth-century Virginians referred to the assent without comment as "the Indians gave the Yo-Hah," indicating it was something known to their audience and without need of explanation. Early in the seventeenth century, Powhatan brought the neighboring Chickahominies into alliance with the Jamestown colonists. The Jamestown leaders went into counsel with the Chickahominy sachems. At the conclusion of negotiations, they sealed the alliance not with a document, but with a speech from each side (published, in the older sense of making the speech before the people). In response, the Powhatan and Chickahominy warriors let out "a general assent and a great shout to confirme it."[38] This served to bind the agreement as signatures would have on a written document.

Perhaps the best description of the assent is Conrad Weiser's 1744 account of treaty-making between the English and the Six Nations. Weiser says:

> When they make treaties with whites, the wholle council and all the warriors perform the "shout of approbation," the "Io——hau." It is performed in the following manner: The Speaker, after a Pause, in a slow tone pronounces the U——huy; all the other Sachems in perfect Silence: So soon as he stops, they all with one Voice, in exact Time, begin one general Io' raising and falling there Voices as the Arch of a Circle, and then raise it as high as the first, and stop at the Height at once, in exact Time; and if it is of great Consequence, the Speaker gives the U——huy thrice, and they make the Shout as often. It is usual, when the white People speak to them, as they give a belt or string of Wampum, for the Interpreter to begin the U——huy, and the Indians to make the Shout.[39]

As late as 1780, A Hessian chaplain noted that Creeks assented to a speech by a British colonel, "as indicated by their mutual Ha!" These vocalizations were not a rhythmic mnemonic device as was often the case with African American antiphony. Nor were they a repetition as in white hymnal antiphony. Instead they served as a communal embodiment of the speech. In effect, Indian antiphony meant that the words were heard and spoken by the whole community. Communal speaking was not merely a metaphor either, as witnessed by a Naudowessie sachem who began a speech by claiming that "I am now about to speak to you with the mouths of these, my brothers, chiefs of the eight bands of the powerful nation of the Naudowessies."[40]

The vocal, nonverbal sound of the assent served as a bridge, linking together the parties in council with each other. The fact that the assent was nonlinguistic was crucial. It bridged communication gaps in situations where all but interpreters found some part of the proceedings unintelligible. Coming only after translation, it was a way of acknowledging across languages and cultures that the other party had been heard and understood.

Treaties and alliances were sometimes chanted into being. In 1685, at a council in Albany between Virginia, New York, and the Five Nations, the Mohawks responded to Virginia's complaints that the Iroquois had not kept up their part of the treaty. In his response to the Virginians, a Mohawk orator "sang all the Covenant Chain over," thus both chanting it back into being and renewing it. He then concluded with a song "by way of Admonition to the Onnondagas, Cayugas and Oneydoes, and concluded all with a Song to the Virginia Indians."[41] In part the song's melody and rhythm may have served as a mnemonic device, but in cultures where identities were sung, the chanting had a much more immediate and active role too.

Another public setting in which the Iroquois sang was the arrival of the French missionaries Joseph Chaumont and Claude Dablon at the Onondagas' main town in 1655. In response to a half-hour speech by one of the missionaries,

> The Chief began the song of response; and all commenced to sing, in wondrous harmony, in a manner somewhat resembling our plain-chant. The first song said that it would take all the rest of the day to thank the Father for so good a speech as he had made them. The second was to congratulate him upon his journey and his arrival. They sang a third time to light him a fire, that he might take possession of it. The fourth song made us all relatives and brothers; the fifth hurled the hatchet into the deepest abyss, in order that peace might reign in all these countries; and the sixth was designed to make the French masters of the river Ontiahantagué. At this point the Captain invited the salmon, brill, and other fish, to leap into our nets, and to fill that river for our service only. He told them they should consider themselves fortunate to end their lives so honorably; named all the fishes of that river, down to the smallest, making a humorous address to each kind; and added a thousand things besides, which excited laughter in all those present. The seventh song pleased us still more, its purpose being to open their hearts, and let us read their joy at our coming. At the close of their songs, they made us a present of two thousand porcelain beads.

While this was not a treaty in European terms, the Onondagas treated it in much the same way as other alliances. To them, spiritual and earthly politics were not necessarily distinguishable, both chanted into being with song.[42]

Wampum and Treaties

First Nations people not only used sounds to present their identities, they used them to imbue things with meanings. The place where this becomes most apparent is in the uses of wampum and pelts in treaty negotiations. There, the punctuation of a speech with a belt of wampum or a beaver pelt rendered the spoken words effective. In turn, the speech entered into the wampum, leaving traces of its meaning. To one skilled in the practices, wampum could be translated to the medium of sound and vice versa. This has often been taken as a form of literacy or money, and there

Fig. 5.1 The belt of wampum delivered by the Indians to William Penn at the "Great Treaty" under the elm tree at Shackamaxon, in 1682. Courtesy of the Library of Congress.

are compelling arguments that such media and the practices around them are like texts, like reading, and like money. From a standpoint originating in Indian worldviews (diverse as they were) rather than European ones, texts, reading, and currency would be understood as like wampum instead of the other way around. By getting beneath the analogy of reading wampum we can arrive at a clearer understanding not of an essential meaning, but of what wampum meant to Native Americans on their terms at particular historical moments.[43]

How reliable are accounts of the use of wampum? After all, they are the work of biased white participant-observers more interested in getting a particular job done than in any modern sense of ethnographic accuracy. It was precisely the jobs they needed to do that give us some reliability in the reports. While they may not have been able to write from a Native American perspective, an approximate understanding of that perspective was necessary for the business at hand to be transacted, whether Jesuit conversions or French, English, and Dutch treaty-making. This was more so the case in the seventeenth century, when Indians often dealt with Whites from a position of power rather than as subalterns.

In order for treaties with the Iroquois to be binding, they had to meet Iroquois standards at some level. At the Albany Congress of 1754, the colonists' speech to the Six Nations was divided into short sections. At the end of each section, looking much like a signature, is the name of the gift to be presented with that section of the speech, usually a string of wampum or a belt. For some semblance of Iroquois ways to be made, there had to first be a ritual address ("brethern . . ."), then a section of a speech given, and a gift presented. This, according to Peter Wraxall, was how to negotiate with the Six Nations without being "obnoxious" to them. Wraxall does not even say it was a satisfactory way, perhaps indicating that he and the colonists felt as if they did not quite understand something of it. Since the negotiations were between state entities rather than simply within them, this system of associating words with things to invest them with meaning must have been spread across all the nations, including native ones, who maintained diplomatic relations with the Iroquois, not just the Iroquois and the English or the French.[44]

What did wampum mean? Daniel Richter points out that for the Iroquois, "gifts made words true." This relationship between words and things was both deeper and wider than that. For many First Nations people besides the Iroquois, a relationship of identity existed between words and things. The present or other thing did not so much stand for the words or sentences; it became their traces. The things embodied the words as visible tracks and made them tangible. Wraxall quotes the Mohawks as saying, "By this belt we desire you to consider what we have said, and by the same we inform you that the Five Nations have something to say to you."[45] A string of wampum and a sentence by themselves were two different things. Only when they were put together in the proper way in a public setting did they come into a relationship of identity. Wampum belts and strings were thus media, the forms utterances took, as well as messages, in the trails they left.

Broad congruencies existed between how different Indians perceived wampum. Otherwise, it would not function as a mediating device between

Fig. 5.2 Naudowessie Indian sachem holding a belt of wampum and speaking to a deer. Jonathan Carver Journal, Add. 8950 folio 169, British Library. Used by permission.

nations with different languages and cultural systems. Like the assent, it had to be capable of crossing cultural and linguistic boundaries. Although an agreement could be made in multiple languages, the tracks of its sounds were stored as an underlying meaning that was not constrained by language, in much the same way Chinese pictographic writing could simultaneously represent mutually unintelligible Chinese dialects using the exact same characters.

Taking this approach, strung or belted wampum, as well as pelts, were like the tracks of a speech event, with the paths of those tracks as well as the content of the speech embodied in the strings and belts. Take for example a drawing that served Carver as a passport up the Chippeway River to Lake Superior. It shows a Naudowessie sachem giving a speech and a belt of wampum to an Ojibwe sachem asking for safe passage from all Ojibwes along the river. The speech sounds of both parties were represented as tracks from the mouth of each to the ears of the other. The drawing and the wampum were the manifestations of those tracks, the visible, tangible effects of the speech sounds. The Jesuit Joseph-François Lafitau claimed that Indians could recognize ethnicities, even particular people, by their tracks. It is likely that wampum marked similar differences, with belts and strings having slightly variant designs and styles from one nation to another. Tracks and tracking played a key role in First Nations life across a wide region.[46] There are many other examples of both voice tracks and animal and human tracks represented in Indian drawings, illustrating their importance. Visually the tracks, whether on the ground or on a page, take a static medium and put it into narrative form, moving through time. The traces of continuous moments left in wampum also moved through time, telling a story rather than recording an event. The visual voice tracks capture the time-constrained aspects of sound in ways that a text might not. The wampum presented embodied the sound, which left its tracks between the two sachems in the drawing. The tracks could be seen and felt as a story in the held belt rather than the speech being "read" from it.

When a band of Delaware Indians returned various stolen shirts and blankets to New Amsterdam city officials in 1656, the officials responded with a speech through an interpreter, saying that Tachpausan, the Indians' sachem, had been wise to return the goods, "for else it might create disharmony and quarrels." In order for the words to have meaning, the New Amsterdam officials sent the Indian messengers back to Tachpausan bearing "a pound of powder" for him that was intended as "a *sign* of our good heart." The Dutch treated it as a sign, with a signifier (the powder) and a signified (the speech). To the Indians this would not fully make sense, for powder was meant to be used up, and the tracks of the speech event could not be seen. Instead it served as a "gift" given in exchange. To overlay anthropological readings of the gift onto Indian belief systems because of European practices distorts those beliefs, perhaps in the process making them seem more familiar and easier to understand from a European-derived frame of reference.[47]

How was wampum used, particularly in relation to sound? Colden described wampum and its use in treaties thus:

> With this [wampum beads], put upon strings, they make these Belts, which they give in all their Treaties, as signs of Confirmation, to remain with the other Party. The Wampum is of two sorts, viz. White and Black; the Black is the rarest, and most valuable. By a regular mixing of the Black and White they distinguish their Belts with various Figures, which they often suit to the Occasion of making use of them.

While Colden understood enough for an outsider, his treatment of wampum as a sign was perhaps an imposition. In contrast, Odianne, a Mohawk orator, referred to a belt of wampum as a "Remembrancer" that would "Stamp Understanding" into its recipients, the other Iroquois, in a council with the English colonists at Albany in 1684.[48] The wampum served to stamp the tracks of the just given-speeches in the minds of the hearers. This is a much more immediate process than Colden's signification accounts for.

Both the Five Nations and the French treated speeches as actions that left tracks in the form of wampum belts. In 1684, a French agent of Governor de La Barre spoke to Garangula, an Onondaga orator. De La Barre threatened to declare war against the Iroquois. He punctuated each statement of his threat with wampum belts and the words *"This Belt Confirms my Words."* In turn, Garangula responded to de La Barre that the Five Nations were not averse to war with France, marking his paragraphs with *"This Belt preserves my Words."* At the end of his reply, he provided a final belt that *"preserves . . . the Authority which the Five Nations have given me"* from speeches at a preparatory council among Iroquois leaders.[49] Perhaps the French had a better understanding of Wampum than the Dutch and Colden, who treated these punctuations as "signs" rather than actions.

George Washington, on a trip to the Ohio River in 1754, called belts of wampum "speech-belts." A Shannoah (Shawnee) sachem, Half King, offered a French speech belt to him, along with other strings and belts, if only he would wait a day. Washington agreed because he "knew that returning of Wampum was the abolishing of Agreements; and giving this up, was shaking off all Dependence upon the French."[50] He already knew the content of the speech that would be given, but he thought it prudent to wait for the giving of the speech out loud in council, accompanied by the appropriate strings and speech belts. Once the tracks of the speech were left, and the French speech, in the form of the speech belt, was turned over, the agreement became documented.

Washington felt it was worth his wait to ensure this accountability. Speaking in a public setting without laying down pelts or belts was the equivalent of trackless speech, the meaning of which would not hold. Three Oneida sachems excused their public slander of the governor during treaty negotiations by saying that "it was said after your Answer, and without laying down either Bever or any Belt or Wampum, as we always do when we make Propositions; Therefore we desire that if it be noted, it may be blotted out, and not made known to Corlaer [the governor]; for we hold firmly to our Covenant, as we said in our Propositions."[51]

One treaty brings together many of the important roles vocal sounds played in these international negotiations. In 1756, Virginia citizens were feeling anxious and vulnerable to the threat of the Catawbas and Cherokees joining the French in the war against the English. Relations with the Indians were already strained on a number of fronts. Catawbas had successfully played internal colonial interests off against each other, so that Virginia had only recently been vying with South Carolina and North Carolina for their loyalties. Virginia had also skipped the unsuccessful Albany Conference, deeming themselves in no need to treat with the Indians on their own terms a few years earlier, yet proceeding themselves on an equally if not more unsuccessful Indian policy of botched aggressions led by George Washington. The Virginians were now anxious to shore things up with the Catawbas and Cherokees, so Virginia sent messengers to both nations to set up treaty councils.[52]

Before arriving, the Virginia delegation had trouble obtaining enough wampum to make the treaty hold. They decided to try to be frugal with it in the hopes that more would be brought, but in the meantime they had to substitute other "presents" instead of the usual number of belts and strings to confirm their speeches. In part this shows a shifting balance of power, as Catawbas and Cherokees had grown more dependent on European goods, whether gunpowder, brandy, or less habit forming stuff. Indian protocols became less rigid. Where a belt or string for each and every proposition would have been required before, by the 1750s, in Virginia at least, a bundle of "goods for a present" punctuated with a belt or string of wampum here and there sufficed.[53]

The speech events that followed were carefully framed, but in the discourse of the Catawbas rather than the Virginians. When the Virginians reached the Catawba Village on February 20, 1756, they began negotiations by reading their commissions aloud, thus verbally establishing their legitimacy to be speaking for all of Virginia. Lieutenant Governor Robert Dinwiddie's instructions concerning his speech reveal interesting notions about Catawba and Virginian beliefs about where sounds came from. He told the commissioners, "(as the Custom of the Indians is) you are to tell them their Brother, the Governor of Virginia, is going to speak to them." The commissioners then read aloud a speech that Dinwiddie had written.[54] Sounds of a voice, for the Catawbas, and it seems for many other First Nations people, did not necessarily belong to the speaker. Groups could speak (as in the deer speaking in Carver's drawing), and one individual could use another's voice without being present. Messengers were said to be the sender speaking.

The sounds were as important as the words in some respects, for the Virginians read Dinwiddie's commission out loud in English, a language that most of the Catawbas did not understand. William Giles, the party's interpreter, then reread the commission in Catawba. The commissioners prepared the way for the governor's written speech to be read aloud by giving a belt of wampum. Heigler, king of the Catawbas, then answered that the ears of the Catawbas were properly ready to hear what the governor had to say. The Virginians read aloud the governor's speech in English, framing it as if it were the governor himself speaking. This was largely a nonverbal act to Catawba ears, so it was next translated into Catawba. Upon its completion, a belt of wampum "confirmed" it, and the Catawbas "*gave the* YO-HAH." The commissioners then gave their own translated talk, punctuated with two strings and one belt of wampum, "*upon which the* Indians *gave the usual Cry of Approbation*" before hearing the translation, thus confirming by assent a speech of which they had not yet heard the content. King Heigler, the leader of the Catawbas, then repeated the speech to the interpreter to ensure its accuracy, taking it to council that evening and promising an answer in the morning, at which time he agreed in substance and "*Gave a Belt of Wampum.*" After that, some warriors spoke out with no wampum presented, indicating that these were just opinions, not binding agreements. The commissioners then responded with a speech, duly translated, and the presentation of their gifts. Finally King Heigler verbally accepted the treaty for the Catawbas. The treaty was then read aloud and translated once more, then signed by the commissioners, King Heigler, and a number of the warriors, the Indians signing with marks. Upon this, the Catawba warriors once more assented with the "YO-HAH." Wine and punch were brought out and a toast was drunk to the King of England and the Catawbas, concluding the treaty.[55] Wampum stamped the meaning of the speeches on the minds of the listeners and the whole event was framed and validated by mutually understood nonverbal cries.

That same year, the government of the colony of Virginia printed the treaty. The primary audience for the printed treaty would be literate people and those they read to in Virginia. The treaty, in the visual medium of print, was meant to be a public document in a public comprising white male property owners. Another audience was interested parties in England, the government, and investors in particular, but also those literate folk seeking knowledge of how things worked in a faraway place. Virginians wanted to show the success of the treaty negotiations. For example, the Virginians addressed the Indians as "brothers" with the English under "our Father the Great King" on "the other Side of the great Water," effectively using patriarchal language to symbolically colonize the Cherokees and the Catawbas, at least to the eyes of the print audience.[56] How did the Catawbas interpret the English frame? From a Catawba frame, the treaty was verified by the conjuncture of words and things—the assents, the treaty speeches, and their trails of wampum. For matrilineal societies, the English discourse of gendered power relations meant little, so the Catawbas and Creeks let it pass, perhaps without comprehension. For them, as for the Iroquois, they answered to these terms "For no other Reason . . . but because he calls us Children. These Names signify nothing." Fathers held no great status; they were considered no more than visitors in the mother's home, perhaps equal to a brother. So while the white Virginians spoke of two equals negotiating under a British superior, the Catawbas heard two equals negotiating, with reference to a distant third party being made.[57] Both parties in the negotiation were probably somewhat aware of the other's meanings and willing to tolerate them, but the Virginians had to follow the protocols of Indian publication by marking each speech with a pelt, a belt, or string of wampum in order for the agreement to have any effect. They then sought to incorporate that frame with the language of gender and the discourse of print, making publication itself seem a colonizing act to the eyes (if not the ears) of the readers.

Memorization

After an Iroquois council had deliberated a proposed treaty and reached consensus, its "resolution is imprinted in the Memory of the One chosen from among them, of great Reputation and Elocution, who is appointed to speak in Publick. He is assisted by a Prompter, who puts him in mind of anything he forgets." Colden reported a similar Iroquois method in 1727:

Fig. 5.3 In this reproduction of a painting by Benjamin West, a warrior seated next to the wampum-holding orator is smoking a calumet, an improbable happening during the speech. Perhaps Grignion or West, not realizing the significance of the prompter's stick in the hands of an elder, changed it into a pipe to fit his own expectations. Courtesy of the Library of Congress.

They commonly repeat over all that has been said to them, before they return any Answer, and one may be surprized at the Exactness of these Repetitions. They take the following Method to assist their Memories: The Sachem, who presides at these Conferences, has a Bundle of small Sticks in his Hand; as soon as the Speaker has finished any one Article of his Speech, this Sachem gives a Stick to another Sachem, who is particularly to remember that Article; and so when another Article is finished, he gives a Stick to another to take Care of that other, and so on. In like Manner when the Speaker answers, each of these has the particular Care of the Answer resolved on to each Article, and prompts the Orator, when his Memory fails him, in the Article committed to his Charge.

This was probably the same method used by the Esopus Indian sachems in complaining to the Dutch in 1659. They "showed 17 staves of wood, with which they signified, that our people had at different places wrongfully beaten and injured their tribe." Communal memorization was still practiced as late as the Revolution on Long Island. Phillipp Waldeck, a Hessian chaplain, described the Long Island Indians' way of keeping track of laws and customs: "Nothing is recorded in writing. When a new law is decreed, the eldest of each family sit together as a court. To each is told what he is to remember, he dare not forget it for fear of death."[58]

Such communal systems of memory operated on principles similar to today's World Wide Web. A body of knowledge on the Web can comprise the "memories" of many different servers. These servers, like individual humans, are "alive" only as long as they have been paid for and kept up—otherwise their contents are gone. A robust network of this type depends on redundant, distributed knowledge, much like the sound-based ones described above. The changing of memories to suit current needs is often seen as a flaw in "oral" cultures. On the Web, however, "continuously updatable content" is thought of as a feature rather than a weakness.

These distributed, redundant networks can survive the loss of any single server or person, recovering data and memories from the built-in redundancy. In the case of a web site, its "mirror" might kick in. In the First Nations example on Long Island, the whole family was responsible for one proposition, so the loss of any member, or even several, would not corrupt the memory. This communal model of memorization is different from the individualistic mnemonic devices to which oral memory is usually attributed (the flaws of "orality" are often pointed out by using the game of "telephone" in which a message is passed around the room from individual to individual with no communal accountability). No doubt rhyming and metric patterns, repetition, and other strategies help individuals remember, but the individualistic nature of literary culture and the modern private act of reading should not blind us to the communal context of Native American remembering.

The decimation of Indian populations through disease and escalated warfare associated with European conquest overwhelmed communal systems of memory. Scaticoke Indians lamented the difficulties of staying "true and faithfull to the thing [an old belt of wampum denoting territorial boundaries]" in light of "our ancient people being almost all dead."[59] In practical terms, the failure of communal memory led to more land loss. This failure was not intrinsic to the medium of storage, though. It was the result of the tremendous trauma incurred upon audible ways of remembering through the loss of vast proportions of the remembering population.

Conclusions

Sound was intimately tied to the project of colonization, but Europeans were not the only powerful, active agents in determining how it was deployed in the seventeenth and eighteenth centuries. Puritans may have been able to linguistically subsume native languages under English in print, but to do so in the world was a much thornier problem that would have to call into play disease and the effects of escalating cycles of warfare to signify anything more than words on a page. Indians and Europeans each took careful stock of how the other sounded. They may have scorned the other's sounds as unmusical, like beasts, as hellish, or as ineffective, but nonetheless they listened and noted with care. European Americans were on Indian lands and at a disadvantage at first, only coming to dominate native peoples as time went on, the process not reaching a peak until the nineteenth century.

European ways of hearing were in many ways more sensitive and powerful than those of the present day, so it is no surprise that they followed the sounds Indians used in battle in great detail. For their part, First Nations people kept these beliefs and practices strong throughout the seventeenth century, but the ravages of constant war and disease strained their ways of using sound to mark and preserve important narratives, ultimately breaking them in many instances, which led to havoc and loss when it came to dealing with land-hungry, ever-growing European American settlers. When no group of Indians could converge upon the meanings and narratives and agreements tracked in a belt of wampum, they were lost, having lost, as it were, their communal memory. But we must not exaggerate this process, for while it happened in a most destructive way, other agreements

that were preserved were run over roughshod by whites once a trend toward dispossession was established. In fact, some of the belts—and their
meanings—remain with Indian nations, their stories intact to this day.
The U.S. government still honors the treaty of Canandaigua, made and
recorded in wampum in 1794, by distributing bolts of cloth to the Six Nations each year. The implications of this treaty for land ownership are still
being played out in the federal court system today, with initial findings in
favor of the Iroquois case.[60] Thus the story is not only one of the break-
down of Indian ways of knowing, but also one of betrayal, domination, and
continuing resistance.

Conclusion: Worlds Chanted into Being

Some sounds, like thunder, physically sounded much the same in early America as they do now. But how they were perceived is an entirely different matter, subject to historical contingencies, and is a matter of historical inquiry. Today, thunder may sound like a loud, scary, but ultimately harmless noise that follows lightning. Until the eighteenth century, however, it was to most North Americans a powerful product of an intelligent being, whether as God's or a demon's voice to Anglo-Americans, as the sonic identity of playful, protecting gods moving in the world to Native Americans, or as an animated and judgmental force to African Americans. Our bells, drums, and fiddles may still sound similar to their seventeenth-century counterparts, but their meanings and social contexts have changed them from important elements of cultural cohesion to merely entertainment. The acoustics of old churches and meetinghouses have changed little if at all, but these spaces are used entirely differently, with electronic amplification saturating them with voice in ways unimaginable to their original designers. Seventeenth-century bells "rang in" English settlers. Native Americans sealed agreements with loud sounds. Enslaved African Americans shifted high-volume sounds banned by slaveholders into quieter ones that helped them build and maintain autonomous cultural spaces under duress. Paralanguage still helps locate people in society, but in a much less potent way than in early America. Those who rant are still on the edge—be it of rationality or of civil discourse—and those who howl would no doubt still be placed altogether outside. Perhaps nothing captures this use of nonverbal sounds better than when a member of an older generation curses a younger generation's music, snarling, "That's not music—you can't even understand the words." Ranting was on the edges of language, but threatened to take over the center with some foreign, demonic tongue rather than being an ineffectual sniping from the

margins. The social power, the sense of threat to civil and ecclesiastical order that the sounds of howls and rants, or even murmurs and whispers,
once caused has grown strange to us.

Rather than thinking of early America as a quieter world, I have sought
to restore the full complexity of its soundscapes. That world had few of the
constant mechanical noises that make the hum and drone of modern life:
no fluorescent lights, refrigerators, fans, or computers. Some mundane
sounds from today would have been impossibly loud then: electric amplification, the roar of jets or loud cars. Yet it is a form of tunnel vision to think
only of modern sounds that did not exist in early America. We have replaced the constant thumping of women's batting staffs with much quieter
washing machines. Street criers, both official and commercial, have been
replaced with the silent visual clutter of street signs, print matter, and bill-
boards. The tremendous clatter of iron shod horse hooves and cart wheels
clacking and grinding against cobblestone roads was louder and sharper
than most automobiles, with their quiet engines and rubber tires on
smooth asphalt streets. Colonial Philadelphia already had noise ordi-
nances in the eighteenth century, and the racket of carts often disturbed
urban church services. Military drums are seldom heard today. Bells rigged
with electric clappers do not project their sounds as far. Many bells have
been replaced altogether with amplified tape recordings of bells, set to
ring at softer volumes so as not to disturb the neighbors who in former
years would be compulsorily moved by their sound.

The questions I have sought to address extend beyond descriptions of
the soundscape. In general, early Americans sensed the world more
through their ears than we do today. If the senses are thought of as filters,
keeping consciousness from being overwhelmed with input, then it follows
that attention to a particular sensory channel could change what McLuhan
called the ratio of the senses. Thus reading and writing, by drawing perception toward vision, would attenuate the auditory. As literacy and
printed matter came closer to saturating North Americans' minds (the
process would not culminate until the early twentieth century), attention
was drawn away from the realm of sound and speech in order to give more
to the visible world.

Native American and African American soundways were much more
complex than a simple attribution of "orality" would allow. Native Americans often thought of sounds as embodied acts of identity. Whereas old
Anglo-American soundways construed thunder as the voice of God, Native
Americans thought of it as a particular act, perhaps unintentional, of a
thunderer. They thought of their own songs and utterances as acts of identity in much the same way they attributed those acts to sounds of thunder
and rustling leaves. In the seventeenth century, Anglo-Americans tended

to depict Native soundways as wild and threatening. Eighteenth-century Anglo-Americans sought to colonize or incorporate Native American soundways into their outworn beliefs from the previous century. In doing so, they translated a First Nations belief in thunder as an unintentional *act* of a thunderer into a belief that the thunder was the *voice* of a god. In changing the Native belief into an outworn European belief, Anglo-Americans could justify colonization by claiming to be "civilizing" ersatz savages. They simultaneously colonized and domesticated Native American soundways, reassuring themselves that Native Americans were developmentally behind.

African Americans used sounds to carve out autonomous agendas in the face of the severe repression of slavery. At first, loud drumming was used as a social enactment of community separate from that of the masters. Masters quickly banned this. Banning drums did not kill off this sonic representation; it merely quieted it. The social functions that African Americans once fulfilled with drums they now carried out with fiddles and other string instruments. They also created community through distinctive styles of call-and-response singing and storytelling. Many of these patterns drifted into Anglo-American life, leading to some creolized cultural elements, such as bluegrass players borrowing an African American instrument, the banjo, and learning how to play it. The cross-cultural interchange grew strange at times, with white banjo players learning songs such as "Bile 'em Cabbage down" and black banjo players learning popular black-face minstrel songs like "Old Dan Tucker."

Seventeenth-century Anglo-Americans carefully structured their soundscapes to create and maintain social order. Communities were laid out in terms of earshot. The failure to do so led to disaster in the Chesapeake in 1622, when settlements spread so far apart that Powhatan Indians were able to successfully divide and attack the settlers. Guns, bells, drums, trumpets, and conch shells were all used to extend this sonic space beyond face-to-face encounters, changing the assumption that "oral cultures" are "face-to-face" societies into the more complex notion that sound could tie together communities across far greater distances than the merely visible.

Seventeenth-century Americans paid careful attention to how public acoustic spaces were made and who had access to a public hearing. Much more care was put into controlling who could say what to whom out loud in public than silently through print. In part, the concern was a function of access. Not everyone could get their hands on a printing press or even write, but all had voices. Thus how and when one spoke was carefully regulated, as was access to the louder instruments and amplifiers, such as church interiors and testers or bells and drums.

The importance of sound reinforces and emphasizes the significance of speech in early America. We have known for some time that gossip, whispering, and murmuring had the power to undermine civil and ecclesiastical order.[1] Understanding the soundways underpinning these social facts makes their importance comprehensible because at that time they were not harmless words. Curses brought about tangible effects in the world. An understanding of physical beliefs about sound's corporeality and agentive nature tells us why, making it a much more adequate explanation than "word magic."[2] The reasons why Puritans and other North American colonists put so much effort into governing speech are underscored with an understanding of the power of sound. Attention to the way they used instrumental sounds to shape and embellish the spaces where properly governed speech took place augments our understanding of their lives: superstitions turn out to be worldviews, compelling even if still foreign to us.

As much as kingdoms, empires, and later nations had to legitimate themselves at the community level through written charters, laws, and proclamations, so too did local communities sound the larger collectivity of the state, colony, or motherland into being through the ritualized ringing of bells and blowing of trumpets and beating of tattoos upon drums. While works by David Cressy, David Waldstreicher, and Len Travers describe the sonic elements of this process and the role of the local community in creating the larger entity, it is only when we understand the efficacy and importance of sound in early American understandings of the world that we can truly explain the process.[3] Thus the nation was a community imagined into being sonically from the bottom up as much as it was visually imagined from the top down through mass print culture. By listening to "dissonant" soundways, such as the "ranting" of Quakers or African American drumming patterns, and "harmonious" soundways, such as the sanctioned use of bells, we begin to understand the many different places from which highly contested notions of American identity emerged. Dissonance and harmony were always a function of who was listening. African American drumming, so troublesome to whites, was a means of creating a covert space that served as a public for African Americans. Bell ringing to protest British policies before the Revolution was dissonant to loyalist ears but harmonious to those with patriotic leanings. These tensions led to plural, contested notions of "American"—which, if we pay attention to soundways, need not be resolved. In light of the nation's subsequent history (it was rent by civil war less than a century after being founded), this continually contested notion of situated American identities—*e pluribus pluribum*—makes more sense than the myth of *e pluribus unum.*

Much of this inquiry has been spent establishing the soundways of early America that stand in contrast to our own. It began, however, by asking how the thunder that was so powerful to Nathaniel Shurtleff's ancestors became lightning to his nineteenth-century readers. From the end of the seventeenth century onward a rapid shift in the ratio of the senses was set in motion that led to soundways that would be much more familiar to a reader today.

This shift was attended by markers of modernization, including increasingly specialized market economies embedded in a growing colonial British empire that reached nearly around the world, and the rise of cheap print in the form of newspapers, broadsides, pamphlets, and local-imprint books. Printing presses, a rarity in the seventeenth century, became increasingly common.

No longer was news the precious commodity carried and held only by those with access to the means of traveling to the sources of the news. It came to each port of the empire with each ship and took on a life of its own. At first it traveled more efficiently by word of mouth through established communication networks attached to colonial markets. Julius Scott has shown that African Americans were highly successful in connecting local communication networks—rumor mills—to ship-borne news carried from port to port by black sailors. T. H. Breen has shown how the network of markets created a common language of goods that he believes arose in the mid-eighteenth century.[4]

News from around the world became a byproduct of empire, increasing in both supply and demand. This led to a loss in its value per item even as its overall value increased as a means of tying various publics together— whether the covert publics of African Americans, the sanctioned publics of empire, or the settler colony publics banding together along the periphery of the empire. Print allowed news to be fixed and distributed as an object, a commodity, along with other commodities.[5] At first, print was the slowest means of distribution and carried only the most staid information. But as presses became more common, printed news took on a life of its own. David Conroy has shown how print and speech were intertwined rather than competitive. Newspapers were read aloud in public houses, combining sonic and visual networks.[6] The visible word grew in importance not by competing with the spoken word but by augmenting it. An understanding of the rise of revolutionary sentiment and the nation that resulted from it needs to take into account the distinctive auditory publics—audiences—into which massive doses of print were injected in the eighteenth century. American historians need to understand the soundways that underlay the emergence of the several eighteenth-century notions of "America."

Sound was not overcome by vision in eighteenth-century America. It is fruitless to say sound is more important than vision or vice versa. Both are necessary components of any culture's perceptual field, and comparing them to establish which was more important is in some ways like comparing apples to oranges. At the same time, however, we have been faced with the inexorable diminution of sound as a force in the early modern world, attended by a rise in the importance of vision. This inquiry has shown that it is possible to quantify the relationship of these two shifts in certain non-absolute ways. We can measure both and establish a relationship between the two sensory channels without ever saying that one is absolutely more or less than the other. McLuhan's ratio of the senses may be thought of as a ratio that results in a complex number. Two numbers from partially incommensurable fields still yield useful results, even when the equation cannot be completely solved.

This inquiry began by marking that shift in the ratio of the senses and asking what it meant. Part of the answer comes from recovering the meaning of the earlier, more sonically oriented world. But why was there a shift? McLuhan's hypothesis predicts that the shift would have taken place much earlier, since moveable type was introduced to Europe before the European exploration of the Americas began. Most historians, however, take no account of the perceptual shift I have documented. They assume that people in the seventeenth century sensed their worlds the same way we do, even if they thought about things differently. So what are we to make of a shift in the senses that took place nearly three hundred years after the invention of movable type, its only ostensible cause? And if early Americans sensed things differently than we do, how are we to understand their worlds if we do not take this shift in the senses into account?

The senses did not shift with the acquisition of literacy on the individual level. The shift toward vision was dependant on what Benedict Anderson and others call mass print culture, in which the effects of literacy are distributed throughout the society, affecting even the illiterate in their pervasiveness. So it was not enough to have a large number of literate people. The society as a whole had to think literately. In order to do this, older ways of thinking had to be disrupted. This was precisely the set of circumstances that native-born colonists faced in the eighteenth century. That was the time when vision came to the fore. It was also the time when the notion of an exceptional American national identity began to take a somewhat protean and contested shape. Anderson argues that the modern form of nationalism arose from the emergence of three necessary conditions: capitalist markets, mass print culture, and the privileging of vernacular languages. Alone, these were not sufficient. They needed to be set in motion by a fourth factor, which Anderson associates with creole identity. Anderson's notion of creole is stunted by his limiting membership in this category to whites. This allows him to associate nation building with print while excluding Europe from the first generation of the process (no *creolité*), while at the same time excluding nonwhites (only whites are creole, and nonwhites have no access to print). Michael Warner goes so far as to say that nonwhites who were literate and had access to print became functionally white.[7]

These exclusions serve to clean up the messy process of disseminating knowledge in early America, but they wallpaper over some very live issues once we pay attention to the soundscape. What if *creolité* cannot in fact be limited to whites? The covert verbal communication networks of African Americans would then need to be accounted for, which a structural analysis of print and capitalism is not capable of alone. Anderson's model of nation building shares a weakness with Bernard Bailyn's account of the same process. Both rely on elite white sources but claim society-wide ideologies as the driving force in the nation-building project.[8] How could these ideologies spread without help from verbal, sonic networks? And if they were in fact dependant on these networks, how is it that the ideological work was all from the top down? If that were the case, the revolution would never have happened.[9] Arbitrarily limiting *creolité* to elite native-born whites is a convenient but nominalistic solution. A truer understanding comes from grappling with the full complexity of early American communication networks. This can only be done once we realize the powerful nature of beliefs about sound that undergirded these networks even to the end of the eighteenth century.

One response has been to claim that no distinctive American identity emerged until after the revolution. Breen finds his evidence in the opening of distinctly American markets only after the Revolution. Carol Smith-Rosenberg, working from popular periodicals, argues that print interpellated the nation into being in the minds of the populace. David Waldstreicher and Len Travers argue that in some ways the idea of the nation was celebrated into being, creating an ideology of Americanism where none existed before.[10] All these stories have merit, but they tend to neglect conditions that allowed the colonists to see and hear themselves as distinctive enough to rebel against England in the first place. *That* notion of Americas is to be found in the contested, pluralistic, polyvocal soundscapes Americans created, maintained, and sometimes fought for in the eighteenth century.

Jürgen Habermas's idea of the public sphere has become influential in explaining the rise of American national identity. He marks a structural transformation in how societies were constructed in the eighteenth century. Capitalism combined with increased literacy, and increased availability of print led to people discussing a uniform set of ideas that they read about in coffee shops, tea houses, and taverns. These locations and the supposed anonymity of the authors led to a leveling of traditional social class that enabled people to discuss ideas solely on the basis of merit. This in turn led to the creation of a public sphere, a shared imagined space where critiques of the state could take place in a neutral fashion, uncoupled from both private interest and state censure by the benefits of laissez-faire capitalism. A historiographical problem with Habermas's idea applied to North America is that his professed project, the defense of the modern liberal nation-state as the best of all worlds, ignores a generation of scholarship that convincingly demonstrates that liberalism was not the driving ideology of the revolution. This is how things ought to be, for Habermas wrote *The Structural Transformation of the Public Sphere: An Inquiry into a Category of Bourgeois Society* in 1962. It was not translated until 1991, at which point the concept of public sphere found a new life.[11]

A single public sphere, to which one was either in—as were landed white free property owners—or out, fails to account for all the grappling over the bounds and content of public acoustic spaces. Political and film theorists have marked multiple overlapping public spheres, but a problem with this approach is that public spheres can no longer do the unifying work they are meant to explain in the first place.[12] Changing the terms from public sphere to audience, earshot, and public hearing gives us a way through the problem. I am not suggesting that sonic terms replace the print- and thus vision-driven notion of public sphere. But attention to soundways richly complicates the cultural context from which the latter arose rather than having it emerge *ab nihilo*. Sounds do not have to be discrete in order to do their work. A dissonant strand gaining access to public hearing might introduce tension, but it does not compete with it. Unlike the visual notion of a public sphere, sounds could occupy the same space without difficulty. Again, soundways give us new ways to think about pluralism and American identities.

Max Weber's explanation for the rise of modernity is that the world became *disenchanted*. We have been considering what we now think of as *enchanted* worlds—that is, worlds chanted into being. Sounds were powerful enough to do that work without resort to the magic we now associate with "enchantment." Cross-cultural treaties required sonic assent or proclamation in order to work. Drumming and bell ringing were considered potent enough that they had to be carefully controlled. It was a world where voiced wills had real effects. Protestant salvation was always already happening—or not—in the lives of individuals. The true voicing of one's heart in a heavenward groan, moan, or cry was the same as deliverance.

But language and reflection subtly undermined this isomorphic act, separating the sounded word from its abstract semantic referent, the deed. The distance induced doubt. If words were language and all of language was what could be reduced to the visible page, then how could one know if her groans were true? The pervasiveness of print culture slowly created a category of "the linguistic" that excludes verbal sounds that could not be reduced to the visible word, or else relegates them to the margins as "paralanguage," locating them with other such dubious enterprises like parapsychology. Doubt about salvation was at the heart of late Puritan jeremiads that some scholars take as a trope defining American identity. It also indicates a shift toward literacy. Groans and wails were as close to the Adamic language as humans could hope for, and yet in a world increasingly textualized, access to one's own heart and the heart of others grew occluded. It was precisely there, in the heart of hearts irreducible to language, that certainty could no longer be had. This was Weber's iron cage. Understanding its profound alienating power requires that we attend to the shift in the senses, something that Weber never considers.

There has been some talk of a "linguistic turn" to history. Actually it is a turn toward the philosophy of language rather than linguistics. It is the product of a visually dominated, hyper-literate epistemology that has mistaken the map for the territory, the text for the world. In it, all the world is treated as "text" to be unraveled and sifted through. This usage of text to indicate styles and discourses that are figuratively "woven" and "read" is a twentieth-century fabrication. Etymologically, "text" derives from the same Latin root as "texture" and "textile," and scholars have used this to extend the domain of texts beyond the written to the world itself, creating discourses with a warp and woof, a textual pattern, and so forth. While this is certainly a valid extension, or perhaps reextension of the text, it must not be confused with the way words—or the world—were thought about in the seventeenth and eighteenth centuries.

From the late fifteenth to the late seventeenth century, text was first and foremost a synonym for the actual words of the printed Bible. The association of a text with biblical authority is well documented from the twelfth century. The first English usages of the word to indicate nonbiblical publications do not appear until late in the fourteenth century. The secularized usage grew more common after the rise of print moved visual publication out of the oligopoly of government and church. "To text" was a transitive verb meaning "to print or write" during the first half of the seventeenth century. The first reference was in 1599, with 1639 marking the last.[13] This evanescent usage shows that "text" was strongly associated with the visual field of print and writing.

In seventeenth-century English, discourse and text stood in a relation-

ship of marked contrast. To twenty-first century academics, the two terms
have nearly, but not quite, converged. Seventeenth-century discourse was
conversation—sounded speech—operating within habitual social parameters. As a verb, it was the precursor of today's "to discuss."[14] Discourse took
place out loud. When it was between heaven and earth or state and community, it took place invisibly. It was the negotiation of meaning via an orderly succession of paralinguistic, verbal, or nonlinguistic vocal sounds. It
was the very process of reason and rationality. These sound-centered usages of the term were obsolete by the end of the eighteenth century. As
the variants concerned made a long slow fade, another meaning of discourse began to gather momentum, becoming primary in the eighteenth
century. This was a reasoned structure of ideas, not necessarily taking place
within the context of a conversation. Again we encounter the dissociation
of the concept from its referent, as discourse ceased to be conversational.
A discourse could be one man's (not gender-neutral) thoughts, silently put
into the stream of the society's conventions by means of print. It was only
once discourse became independent of conversational sound that the
world could be monologically treated as a text. Engaging in conversation
became merely to discuss, with the connotation of reasoning lost. This
process parallels the diminution of the sound of thunder as the powerful
element in storms. This, then, was the disenchantment of the world. The
scholarly metaphor for life shifted from a conversation to a book. Only
then could vision colonize the domain of reason.

  Listening to early American discourses of hearing with an awareness of
visual, text-centered habits in mind is an exercise that does more than letting us imagine past worlds and how they became our own. Text-based visual ways of knowing are in a state of flux today, much like they were at the
end of the seventeenth century. Hypertext, and the various audio and
video recording and transmission technologies, are shifting the print-
based ground over which scholars traditionally walk. New, disorienting
ways of knowing are creeping in, and some old ways are back, disguised as
new. Without admitting that our ways of knowing are habits rather than
facts, there is no way to respond. This was McLuhan's point—"uncon-
sciousness of any force is a disaster, especially one we have made ourselves."[15] We need not be victims of our own technologies, but unless we
recognize shifts in perceptual habits that technology engenders over time,
we can be nothing but the victims, never knowing what struck us, like the
deer too terrified of the headlights or the horn to respond to the truck.
Hardly technological determinism.

  If history is a conversation with the past, what does a history of the audible world have to tell us today? It took centuries to unlock the transfor-

mative potential of print. Seventeenth-century print culture was integrally linked to the audible world from which it came. During the eighteenth-century shift toward visually centered ways of thinking, a great deal of confusion arose around the regulation of old and new ways of representation.

We are in a similar position today with the advent of hypertext. Like writing and print, hypertext extends the possibilities of representation by the introduction of a simple function unrelated to language. Readers rather than authors control the structuring of what is read. As a result, readers used to linear texts get disoriented by the absence of discernible authors. Designers respond by trying to make hypertext systems more familiar-looking, which misses the point. The present-day reaction to the Internet, which includes everything from prideful ignorance to millennial waxing about the global village to come, will probably seem as much in need of a translation as Shurtleff's thunderstorms in a few years.

Presently, if pundits, experts, and the computer industry are to be believed, we rest on the brink of a transformation precipitated by the development of hypertext and cyberspace. Like literacy in early print culture, there are competing, inchoate models of cyberliteracy (hypertext exists, but no one is sure how to read it yet). Often, they are reworkings of familiar media ("virtual magazines" are the most obvious example). The new media seem disorienting to many, the harbingers of the end of rationality for others—much as early critics of the printed word warned of the decay of the spoken arts and aural modes of thought in the face of the kudzu-like expansion of printed matter. The digiterati blithely dismiss such warnings and continue into the new perceptual shift toward multiple, windowed, virtual realities. Often these new worlds look suspiciously like typing. Like the scribal milieu that long dominated the first century's output of print, the evolution of cyberspace is bound to give up its secrets and realize its potential only slowly to people so long ensconced in older media. Are new media leading to a shift in the ratio of the senses today? Quite possibly, but that shift may be of the same slow, unconscious sort that marked the changes that attended modernization in the eighteenth and nineteenth centuries.

If there is to be another shift, what will it look like? Two important features of the latest media revolution seem curiously familiar. Hypertext is touted as being continuously updatable and massively parallel. First, continuous updatability means that texts are no longer fixed and thus need never become obsolete. Errors can be corrected. But the fixity of knowledge was what was touted as the wellspring of civilization in the literacy theory. It allowed thoughts to be concretized well enough that they could stay put long enough for ever more intricate and abstract knowledge to be formed. What was a flaw of "oral" cultures—for they were nothing if not continuously updatable—has become a feature of hyperliteracy. Second, wide-area networked computers, with the Internet as the prime example, are connected by many different routes, so that if any one node is knocked out, the network stays connected and no knowledge is lost. This massive redundancy is the measure of a "robust" network. Knowledge is distributed rather than centralized in the new media. But the distribution of knowledge across a human network was precisely the "flaw" that print and literacy corrected through the process of centralizing knowledge into authoritative editions. In principle, massive redundancy is no different from Native American communal memorization.

With this ambiguous, lumbering sort of post-modern transformation in mind, it becomes particularly relevant to dispense with the notion of "oral culture" as some kind of primitive state of being. Such a position is only tenable if one holds that text-based ways of thinking are an ahistorical, natural predecessor to whatever electronic mode of consciousness is upon us (but not clearly visible) today. Though we can see through the flux of the present no better than early modern Europeans could, a historical study of attitudes and beliefs about sound in the seventeenth and eighteenth centuries can help us come to terms with our present unavoidable naivete about the newest of technological revolutions. The persistence of the age of the ear even to our own day might provide more than solace. A study of the legitimacy of the soundscape that was antiquated by the rise of vision can help us understand why we tend to read about the information revolution in magazines rather than "hyper-reading" about it in cyberspace.

I sit in a room visually cloistered from the outside world to keep the glare off my computer screen as I type. In some respects, my computer is a hypervisual environment. I am working with texts on a silent, luminous screen, with just the hum of the machine's fan filling out the soundscape. I never see the flash of lightning from the unseasonable rainstorm, but I hear the thunder and know from experience that I must unplug the computer or risk losing all my work as I translate my hypertext notes into a linear, printable narrative. I am reduced to paper, pen, and books if the lights are not knocked out, leaving me in darkness. If that happens, it is the sound of conversation alone to sustain us.

Acknowledgments

T his book could not have been completed without the support—material, emotional, and intellectual—I received from my mentors, friends, and family, as well as the institutions with which I have been affiliated. I have had generous financial support from a Crown fellowship at Brandeis University. A dissertation fellowship from Brandeis allowed me to deepen my research extensively, and a last-minute grant from the Department of History bridged a critical gap that allowed me to finish writing. A fellowship at the John Carter Brown Library in Providence in 1997 was a tremendous boon both for research and camaraderie. Karen Graubart and David Murray stand out as friends who have helped me think about this project in new ways. A fellowship at the Smithsonian Institution provided me with the flexibility and time to think through parts of this work. While there, Pete Daniel and Peter Seitel both acted as particularly friendly and thoughtful sounding boards. A Gest Fellowship at Haverford College's Quaker Collection provided a wonderful opportunity to research Quaker material culture and soundways; everyone there made my short stay wonderfully productive. A Mellon Fellowship at the Virginia Historical Society was particularly useful in working through Southern soundways. A Peterson Fellowship at the American Antiquarian Society allowed me to work with the collection's tremendous manuscript resources and a vast number of local histories. Finally, a fellowship from the Institute for the Advanced Study of Religion at Yale enabled me to more carefully document the acoustics of meetinghouses and churches. It also gave me the time to complete the process of turning a dissertation into a book.

The editors and staff at Cornell University Press have been simultaneously supportive and efficient, enabling this book to come to press quickly. I am particularly grateful for the work and occasional patience of Sheri

Englund and Nicole Hanley. In addition, the anonymous readers provided excellent constructive comments.

David Hackett Fischer has put a tremendous investment of time and effort into seeing this book completed. His close critical readings and demand that the project bear on important questions in American history are just two ways in which he has strengthened this work. His optimism, his belief in me, and his unstinting support from inception to completion have made this book possible. Thanks!

Jane Kamensky provided incisive comments that have greatly strengthened the book. James Kloppenberg poked and prodded at the manuscript's weakest structural points, making me think carefully about the task of historical scholarship. Ibrahim Sundiata has been a great friend and a lively mind, ever skeptical and unfailingly supportive at the same time. My conversations with Jacqueline Jones, Judith Irvine, and Steven Harris, and Karen Kupperman all turn up in the book in one way or another, in ways I hope they will approve. Chris Warren, Ann Plane, Mark Taylor, Mark Renella, and Holly Snyder have been wonderful colleagues at Brandeis and beyond. John Brooke's comments were crucial to shaping the conceptual frame of this project. He has been an incisive and supportive reader. My colleagues and students at Oberlin College, Hamilton College, New York University, and the University of Hawai'i at Mānoa have made me think through parts of this project much more clearly than I would have otherwise.

Many parts of this work began as papers in front of lively and critical audiences. The John Carter Brown Library's luncheon lectures gave me the opportunity to try out my ideas about New England's ranters. The Massachusetts Historical Society provided a fabulous public hearing for a later version of the same paper, giving me detailed and constructive feedback in a very lively conversation. Particular thanks go to Steve Marini of Wellesley College for his close reading of the paper. The chapters on the natural soundscape were presented to the Society for Early America. The Chesapeake materials benefited greatly from the comments I received from Phil Morgan, James Axtell, Robert Gross, Ron Hoffman, and others at the Omohundro Institute of Early American History and Culture. Sections on African soundways were presented as job talks at Oberlin and Hamilton Colleges, at the Southern Historical Society meeting in Atlanta in 1992, at a conference on creolization at the University of Delaware in 1995, Northeastern University's World History Association Seminars, and at the annual meeting of the Linguistic Society of America in 1992. I gave a multimedia presentation of the chapters on material culture at the annual meeting of the American Historical Association in 1999. All these presentations allowed me to see the material in new ways that I hope have improved the book.

John Osborne, John Thornton, Francis Bremer, and Dennis Downey provided me with excellent undergraduate training in history at Millersville University. I still rely on their mentorship, whether they know it or not. I also owe an unusual set of debts as a nontraditional scholar. Kevin Anderson and Rich Blythe employed me as a cook for several years as I worked my way through my undergraduate degree. Without their support and their occasional tolerance for lackadaisical performance and lateness around a semester's crunch times, this work would never have even gotten started.

Portions of the Introduction and Chapter 2 have appeared in Richard Cullen Rath, "African Music in Seventeenth-Century Jamaica: Cultural Transit and Transition," *William and Mary Quarterly* 50, no. 3 (1993): 700–726, and are used with permission. Parts of Chapter 2 also appeared in Richard Cullen Rath, "Drums and Power: Ways of Creolizing Music in Coastal South Carolina and Georgia, 1730–1790," in *Creolization in the Americas: Cultural Adaptations to the New World,* ed. Steven Reinhardt and David Buisseret (Arlington, Tex., 2000). They appear courtesy of Texas A&M Press.

I owe thanks to Gwen Agina and Susan Abrams in the Department of History at the University of Hawai'i at Mānoa for their administrative help. Also, the librarians at Brandeis University, Oberlin College, Hamilton College, Syracuse University, and the University of Hawai'i have all been tremendously helpful and, at times, patient and tolerant above and beyond the call of duty.

Ruchira Das Gupta, Ed Young, Karen Vasudavan, and Serena Sundaram have each contributed to making the completion of the book possible with their love and friendship. Our friends in Boston, New York, Syracuse, and Hawai'i have been a wonderful and sustaining network. Thanks in particular go to Riti Sachdeva, Laurie Prendergast, Ramani Sripida-Vaz, Danny Sripada-Vaz, Kalpana Neitzsche, Sadhana Bery, Lyn Bigelow, Sudipta Sen, and Becky Thompson. Karabi Das Gupta informed her daughter Monisha that the project would in fact be completed some day. I am glad that the elder Ms. Das Gupta, along with David Fischer, was sure. She is sorely missed. Arun Das Gupta was supportive in his own way. My parents and siblings have all pitched in at various points: thanks, Mom, Dad, Mark, Joan, Anne, Charla, and Barbara.

Closer to home, Monisha Das Gupta has been a constant companion and support. She has discussed every twist and turn of this project with me, a few of which have been somewhat surreal. At certain crucial junctures, it was Monisha's efforts that carried the day, as she sometimes sacrificed her own work to make it possible for me to bring this book to completion. With much love, I dedicate the book to her.

Abbreviations

AAS	American Antiquarian Society
CCHS	*Collections of the Connecticut Historical Society*
CMHS	*Collections of the Massachusetts Historical Society*
CWCJS	Philip L. Barbour, ed., *The Complete Works of Captain John Smith (1580–1631)*, 3 vols. (Chapel Hill, N.C., 1986)
DRCHNY	John Romeyn Brodhead, Berthold Fernow, and E. B. O'Callaghan, *Documents Relative to the Colonial History of the State of New-York*, 15 vols. (Albany: Weed Parsons, 1853–1887)
H5N	Cadwallader Colden, *The History of the Five Indian Nations Depending on the Province of New-York in America* (New York, 1727)
JCB	John Carter Brown Library
JR	Reuben Gold Thwaites, *The Jesuit Relations and Allied Documents: Travels and Explorations of the Jesuit Missionaries in New France, 1610–1791*, 73 vols. (Cleveland, 1896–1901)
OED Online	*Oxford English Dictionary*, online edition <http://www.oed.com/>

Notes

Introduction

1. Nathaniel Shurtleff, *Thunder and Lightning* (Boston, 1850), i, 5–8, 10, 14, 37, 38, 51 (emphasis mine).

2. Sidney Perley, *Historic Storms of New England* (Salem, Mass., 1891), 68; *JR*, 10: 194–97, 58: 278–79.

3. My use of *ways* in cultural history is discussed in Richard Cullen Rath, "Drums and Power: Ways of Creolizing Music in Coastal South Carolina and Georgia, 1730–1790," in *Creolization in the Americas: Cultural Adaptations to the New World*, ed. Steven Reinhardt and David Buisseret (Arlington, Tex., 2000), 99–130. The call for a history of the senses originated in the 1920s with Lucien Paul Victor Febvre, "A New Kind of History," in *A New Kind of History: From the Writings of Febvre*, ed. Peter Burke (London, 1973). Also see Alain Corbin, *Time, Desire, and Horror: Towards a History of the Senses* (Cambridge, 1995); Constance Classen, *Worlds of Sense: Exploring the Senses in History and Across Cultures* (London and New York, 1993); David Howes, *The Varieties of Sensory Experience: A Sourcebook in the Anthropology of the Senses* (Toronto, 1991). For the history of sound, see Mark M. Smith, *Listening to Nineteenth-Century America* (Chapel Hill, N.C., 2001), Leigh Eric Schmidt, *Hearing Things: Religion, Illusion, and the American Enlightenment* (Cambridge, Mass., 2000); Bruce R. Smith, *The Acoustic World of Early Modern England: Attending to the O-factor* (Chicago, 1999); Alain Corbin, *Village Bells: Sound and Meaning in the Nineteenth-Century French Countryside* (New York, 1998); Emily Ann Thompson, "'Mysteries of the Acoustic': Architectural Acoustics in America, 1800–1932" (Ph.D. diss., Princeton University, 1992).

4. Marshall McLuhan, *The Gutenberg Galaxy: The Making of Typographic Man* (New York, 1962), 124, 183, 249. I discuss the literature on orality and literacy extensively in "Orality, Literacy, and Shifts in Perception in Early American History," http://way.net/rcr/orality.html.

5. McLuhan, *Gutenberg Galaxy*; Marshall McLuhan, *Understanding Media: The Extensions of Man* (New York, 1964); Walter J. Ong, *The Presence of the Word: Some Prolegomena for Cultural and Religious History* (New York, 1970); Walter J. Ong, *Orality and Literacy: The Technologizing of the Word* (London, 1982); Lucien Paul Victor Febvre, *The Problem of Unbelief in the Sixteenth Century: The Religion of Rabelais* (Cambridge, Mass., 1982), 388–90; Jack Goody and Ian Watt, "The Consequences of Literacy," *Comparative Studies in Society and History* 5, no. 3 (1963): 304–45; Jack Goody, *The Domestication of the Savage Mind* (Cambridge, 1977); Jack Goody, *The Interface between the Written and the Oral* (Cambridge, 1987).

6. James Axtell, *The Invasion Within: The Contest of Cultures in Colonial North America* (Ox-

ford, 1985), 14–15, 338 n. 27 (which cites Ong and Goody); Daniel K. Richter, *The Ordeal of the Longhouse: The Peoples of the Iroquois League in the Era of European Colonization* (Chapel Hill, N.C., 1992), 46–47, 307–8 n. 36 (which cites Axtell); Michael G. Kammen, *Colonial New York: A History* (New York, 1975), 96–97, 121–22; Rhys Isaac, *The Transformation of Virginia, 1740–1790* (Chapel Hill, N.C., 1983), 98, 121–31, 373 n. 12, 377 n. 8; Linda K. Kerber, *Women of the Republic: Intellect and Ideology in Revolutionary America* (Chapel Hill, N.C., 1980), 190; Carol F. Karlsen, *The Devil in the Shape of a Woman: Witchcraft in Colonial New England* (New York, 1987), 268–69 n. 10; Richard D. Brown, *Knowledge Is Power: The Diffusion of Information in Early America, 1700–1865* (New York, 1989); Michael Warner, *The Letters of the Republic: Publication and the Public Sphere in Eighteenth-Century America* (Cambridge, Mass., 1990), 11–12, 27–28. Brown and Warner both rely on Isaac, who builds his notion of orality from Ong and Goody.

7. For a concise review and criticism of the notion of the oral/primitive/savage mind and its contrastive association with literacy and "civilization," particularly as it relates to colonial North America, see Francis Jennings, *The Invasion of America: Indians, Colonialism, and the Cant of Conquest* (New York, 1975), 6–12, 15–16.

8. John Smith, *True Relation,* in *CWCJS,* 1: 59.

9. David D. Hall, ed., *Witch-Hunting in Seventeenth-Century New England: A Documentary History, 1638–1692* (Boston, 1991), 135–45, quotes on 136, 140, 145.

10. Isaac, *Transformation of Virginia, 1740–1790;* Kathleen M. Brown, *Good Wives, Nasty Wenches, and Anxious Patriarchs: Gender, Race and Power in Colonial Virginia* (Chapel Hill, N.C., 1996); Brown, *Knowledge Is Power;* Christopher Grasso, *A Speaking Aristocracy: Transforming Public Discourse in Eighteenth-Century Connecticut* (Chapel Hill, N.C., 1999); Jane Neill Kamensky, *Governing the Tongue: The Politics of Speech in Early New England* (Oxford, 1998); Robert Blair St. George, "'Heated'" Speech and Literacy in Seventeenth-Century New England," in *Seventeenth-Century New England,* ed. David G. Allen and David D. Hall (Boston, 1985), 275–322; D. R. Woolf, "Speech, Text, and Time: The Sense of Hearing and the Sense of Past in Renaissance England," *Albion* 18, no. 2 (1986): 159–93.

11. David D. Hall, *The Antinomian Controversy, 1636–1638: A Documentary History* (Middletown, Conn., 1968), 243–44, 246, 263; Nathaniel Bradstreet Shurtleff, ed., *Records of the Governor and Company of the Massachusetts Bay,* 5 vols. (Boston, 1853), 1: 207 n, 301 n, 211–12.

12. Shurtleff, *Records of the Governor and Company,* 1: 406.

13. Richard Allestree, *The Whole Duty of Man* (1657; reprint, Williamsburg, Va., 1746), 255–56.

14. Isaac, *Transformation of Virginia,* 98, 121–31, 373 n. 12, 377 n. 8.

15. *JR,* 33: 25.

16. Hans Sloane, *A Voyage to the Islands Madera, Barbados, Nieves, S. Christophers and Jamaica* (London, 1707), 1: xlvii. For a fuller treatment of these songs, see Richard Cullen Rath, "African Music in Seventeenth-Century Jamaica: Cultural Transit and Transition," *William and Mary Quarterly* 50, no. 3 (1993): 700–726.

17. John Thornton, personal communication, March 1, 1993; Hazel Carter, personal communication, March 19, 1993; David Dalby, "Ashanti Survivals in the Language and Traditions of the Windward Maroons of Jamaica," *African Language Studies,* 12 (1971): 41, 45–46.

18. L. P. Hartley, *The Go-Between* (London, 1953).

Chapter 1. "Those Thunders, Those Roarings": The Natural Soundscape

1. Simon Harward, *A Discourse of Lightnings* (New York, 1971).

2. Harward, *Discourse of Lightnings,* sig. C3r. Seventeenth-century Jesuits writing in North America corroborated this theory. Jacques Marquette wrote that "Noisome vapors arise Con-

stantly arise" from the "mud and mire" of Green Bay, thus "Causing the loudest and most continual Thunder that I have ever heard." Claude Allouez confirmed Marquette's observation, claiming that Green Bay was a great swamp that gave rise to "loud and frequent Peals of thunder." *JR*, 59: 99; *JR*, 60: 161. On bells and other instruments of sound dispelling thunderstorms, see Chapter 2.

3. William Strachey, *The Historie of Travell into Virginia Britania* (London, 1953); William Strachey, *A True Reportory of the Wreck and Redemption of Sir Thomas Gates, Knight, upon and from the Islands of Bermudas*, in *A Voyage to Virginia in 1609*, ed. Louis B. Wright (Charlottesville, 1964); Sylvester Jourdain, *A Discovery of the Bermudas, Otherwise Called the Isle of Devils*, in Wright, *Voyage to Virginia*. For speculation on the identity of the woman to whom Strachey's letter was addressed, see Wright, *Voyage to Virginia*, x–xi.

4. Harward, *Discourse of Lightnings*, sig. C3r–v.

5. Ibid., sig. C2v. Biblical citations given as Psalms 77:17, Psalms 18:13, and Job 38:25. Some may protest that the older Hebrew texts might have used a word other than thunder (though that is improbable). Even if they did, though, the use of thunder in the English translation would be the better indicator of English-derived soundways. The supporting evidence is in the use in the soundways rather than in the origin.

6. Harward, *Discourse of Lightnings*, sig. B1r–B2r.

7. J. L. Austin, *How to Do Things with Words* (Cambridge, Mass., 1962); John R. Searle, *Speech Acts: An Essay in the Philosophy of Language* (London, 1969). In feminist and postmodern readings of speech act theory, hate speech and inequitably gendered language are performative and perlocutionary, broadening once again the scope of what it means to do things with words. See Jane Neill Kamensky, *Governing the Tongue: The Politics of Speech in Early New England* (Oxford, 1998), 191–93.

8. Wright, *Voyage to Virginia*, xx–xxvi; Philip L. Barbour, ed., *The Jamestown Voyages under the First Charter, 1606–1609*, 2 vols. (London, 1969), 2: 249–53; *CWCJS*, 1: 121–30. For the tenuous nature of early settlers' relations with the Indians, see Karen Ordahl Kupperman, *Indians and English: Facing off in Early America* (Ithaca, 2000).

9. Strachey, *True Reportory*, 4, 6, 20; Gabriel Archer to an unknown friend, Virginia, August 31, 1609, in Barbour, *Jamestown Voyages*, 2: 281.

10. Strachey, *True Reportory*, 4, 5.

11. Ibid., 4, 14; John Smith, *An Accidence or the Pathway to Experience (1626)*, in *CWCJS*, 3: 21; and John Smith, *A Sea Grammar (1627)*, in *CWCJS*, 3: 92; Archer to friend, 2: 281. On signals as the admiral speaking, see anonymous, "Day, Night, and Fog Signals Decoded: A Book Used in the Royal Navy at the Time of the Command of Admiral Sir Edward Boscawen," MS, John Carter Brown Library, Codex Eng 32, c. 1747, 12 verso, 18 recto, 23 verso; Smith, *Accidence*, 3: 23; and Smith, *Sea Grammar*, 3: 103. Eighteenth-century signal books are much more visual in their emphasis. Compare "Day, Night, and Fog Signals" and Nicholas Charrington, "Signals to be Observed in Sailing and Fighting by his Majesty's Squadron on the Leeward Island Station," MS, John Carter Brown Library, Codex Eng 72, 1778, with Smith's accounts of signaling cited above.

12. Strachey, *True Reportory*, 95–96; John More to William Trumbull, London, November 9, 1609, in Barbour, *Jamestown Voyages*, 2: 285.

13. For the first Jamestown voyage see George Percy, "Observations Gathered out of a Discourse of the Plantation of the Southerne Colonie in Virginia by the English, 1606," in Barbour, *Jamestown Voyages*, 1: 133. Raleigh is quoted in Charles Norman, *The Discoverers of America* (New York, 1968), 165. For Cuba, see Cabeza de Vaca, *Adventures in the Unknown Interior of America* (Albuquerque, 1983). For other storms during the age of discovery, though with no discussion of their sounds, see David McWilliams Ludlum, *Early American Hurricanes, 1492–1870* (Boston, 1963), 1–10.

14. Robert Beverly, *The History and Present State of Virginia* (Chapel Hill, N.C., 1947).

15. For a list of all the titles, see Richard Cullen Rath, "Worlds Chanted into Being: Soundways in Early America" (Ph.D. diss., Brandeis University, 2001), Appendix I.

16. Robert Dingley, *Vox Coeli, Or, Philosophical, Historicall, and Theological Observations of Thunder* (London, 1658); Robert Barclay, *The Anarchy of the Ranters, and Other Libertines, the Hierarchy of the Romanists and Other Pretended Churches, Equally Refused, and Refuted, in a Two-fold Apology for the Church and People of God Called in Derision Quakers* (1696; reprint, London, 1771), 5–6; *A Thunder-Clap to the Army and Their Friends Warning Them of Their Imminent Danger, and Awakening Them to a Speedy Prevention* ([London], 1648); John Rogers, *Sagrir, or, Doomes-Day Drawing Nigh, with Thunder and Lightening to Lawyers in an Alarum for the New Laws, and the Peoples Liberties from the Norman and Babylonian Yokes* (London, 1654).

17. Edward Burrough, *A Trumpet of the Lord Sounded out of Sion* (London, 1656); Humphery Bache, *The Voice of Thunder, or the Sound of a Trumpet Giving a Certain Sound* (London, 1659), 1; R. S., *The Dreadful and Terrible Voice of God Uttered from the Throne of his Justice as the Voice of a Mighty Thunder, and as the Voice of many Waters Rumbling*, Haverford College Quaker Collection, broadside box ([London, 1660?]).

18. Dorothy White, *An Alarm Sounded to Englands Inhabitants but Most Especially to Englands Rulers* (London, 1661), 1, 3, 4, 5, 7; Dorothy White, *A Lamentation unto This Nation: and Also a Warning to All People of This Present Age and Generation with the Voice of Thunder Sounded Forth from the Throne of the Lord God: and This Is More Particularly a Warning unto the Inhabitants of England and Is to Go Abroad Thorow All Parts of this Nation* (London, 1660).

19. "Boanerges" is often translated as "sons of thunder," a nickname Jesus gave to James and John, the sons of Zebedee, in Mark 3:17. The word is probably derived from the Aramaic compound of the word for "son" and the word denoting "to tremble, rage, or quake." James Strong, *The New Strong's Exhaustive Concordance of the Bible* (Nashville, 1997), s.v. 993, 1121, 1123, 7264–66. That thunder is important in the King James Version but not in the original texts underscores rather than undercuts the argument being made here. John Jewel, *Dad Seiniad Meibion Y Daran Sef Ail-printiad O Lyfr Escob Juel a Elwir Deffyniad Ffydd Eglwys Loegr: Ac O Epistol Yr Escob Dafies at Y Cembru* = an Eccho of the Sons of Thunder, Being a Second Impression of Bishop Juel's Apologie, and of Bishop Davies His Epistle, in the British Tongue (Rhydychen, 1671); *Boanerges, or, The Parliament of Thunder with Their Sharp Declarations against Those Rebellious Malignants That Revolt and Fall Away Form Them for Which They Are Voted Evill Members of the Common-wealth* (London, 1643); Henry Adis, *A Fannaticks Alarm, Given to the Mayor in His Quarters, by One of the Sons of Zion, Become Boanerges to Thunder out the Judgements of God against Oppression and Oppressors* (London, 1661); Samuel Chidley, *Thunder from the Throne of God against the Temples of Idols* ([London], 1653); Edward Burrough, *The Memorable Works of a Son of Thunder and Consolation* ([London], 1672).

20. William Bayly, *Seven Thunders Uttering Their Voices, and the Seven Last Trumpets Preparing to Sound in the Spirit of Jealousies* ([London], 1665); Jane Lead, *The Revelation of Revelations Particularly as an Essay towards the Unsealing, Opening and Discovering the Seven Seals, the Seven Thunders, and the New-Jerusalem State* (London, 1683); *A Calendar of Prophetick Time, Drawn by an Express Scripture-Line* (London, 1684).

21. *A Strange and Wonderful Relation of a Clap of Thunder Which Lately Set Fire to the Dwelling-house of One Widow Rosingrean* (London, n.d.); *A Full and True Relation of the Death and Slaughter of a Man and His Son at Plough, Together with Four Horses* (London, 1680); *A True Relation of the Sad and Deplorable Condition of a Poor Woman in Rosemary-lane near Tower-hill . . . Together with an Account of a Man Living near Shorditch Church, Who Was Struck with the Thunder or Lightning on Tuesday May the 18th* (London, 1680).

22. Increase Mather, *An Essay for the Recording of Illustrious Providences* (Boston, 1684), 72.

23. Ibid., 91–92, 109, 110.

24. Ibid., 7r–8v, of the unnumbered preface.

25. Nathaniel Morton, *New England's Memorial* (Boston, 1855), 55 (emphasis in original);

Mather, *Essay for the Recording of Illustrious Providences,* 73–74; Nathaniel Bradstreet Shurtleff, *Thunder and Lightning; and Deaths at Marshfield in 1658 and 1666* (Boston, 1850), 22.

26. Increase Mather, diary entry for July 16, 1665, AAS, Mather Family Papers, Increase Mather, Box 3, Folder 1 (typescript by Michael G. Hall); George Keith, *The Presbyterian and Independent Visible Churches in New England and Elsewhere Brought to the Test* (London, 1691), 168, 226.

27. "Manuscript of Nathaniel Thomas, in Rev. Samuel Arnold's letter to Increase Mather," cited in Shurtleff, *Thunder and Lightning,* 18. Thomas was apparently at the scene when the lightning struck. William Jones to Increase Mather, 1682, in "Mather Papers," *CMHS,* 4th series, 8: 612–13.

28. Mather, *Illustrious Providences,* 74, 76, 78, 80–84; Shurtleff, *Thunder and Lightning,* 22–24.

29. Increase Mather, *The Day of Trouble Is Near* (Boston, 1674), 25–27. See also Michael G. Hall, *The Last American Puritan: The Life of Increase Mather, 1639–1723* (Middletown, Conn., 1988), 97–100, esp. 99.

30. Mather, *Essay for the Recording of Illustrious Providences,* unnumbered preface, 109, 124–26, 130; Keith Thomas, *Religion and the Decline of Magic: Studies in Popular Beliefs in Sixteenth and Seventeenth-century England* (New York, 1971), 94–95, 95 n. 1.

31. Mather, *Essay for the Recording of Illustrious Providences,* 109, 116–18.

32. Charles Morton, "Compendium Physicae," *Publications of the Colonial Society of Massachusetts* 33 (1940): 85–87; Charles Morton, "Compendium Physicae," AAS, Mss. Dept., Octavo vols. "M," c. 1700, 98–99, 102. The "Compendium" was not published until 1940. Students wrote manuscript versions of Morton's lectures. There are twenty-four versions extant, dated from 1679–80, when it first appeared, and 1738, by which time it was somewhat outdated. I am grateful to Thomas Knowles, the manuscript archivist at the American Antiquarian Society, for sharing his knowledge of the "Compendium" with me.

33. Francis J. Bremer, "Increase Mather's Friends: The Trans-Atlantic Congregational Network of the Seventeenth Century," *Proceedings of the American Antiquarian Society* 94, no. 1 (1984): 59–96; Francis J. Bremer, *Shaping New Englands: Puritan Clergymen in Seventeenth-Century England and New England* (New York, 1994).

34. Morton, *New England's Memorial,* 139; "Roger Williams to John Winthrop," *CMHS,* 6: 229; Mather, *Essay for the Recording of Illustrious Providences,* 322.

35. Samuel Sewall, *Letter-Book of Samuel Sewall, CMHS* 6th series, vol. 1–2 (Boston, 1886–88), 2:232, cited in David D. Hall, *Worlds of Wonder, Days of Judgment: Popular Religious Belief in Early New England* (Cambridge, Mass., 1990), 220.

36. David P. Hill, "The Sound of an Earthquake," *Earthquake Information Bulletin* 8, no. 3 (May–June 1976): 15–18; Joshua Moodey to Samuel Nowell, December 15, 1688, in Mather Papers, *CMHS,* 4th Series 8 (1868): 372.

37. On the Moodus noises in general, see Carl W. Stover and Jerry L. Coffman, *Seismicity of the United States (Revised),* U.S. Geological Survey Professional Paper 1527 (Washington, D.C., 1993); Carla Helfferich, "Things That Go Boom in the Night," *Alaska Science Forum,* Article 896 (19 October 1988): <http://www.gi.alaska.edu/ScienceForum/ASF8/896.html>. Also see David D. Field, *A Statistical Account of the County of Middlesex, in Connecticut* (Middletown, 1819), 62–63, and David D. Field, *History of the Towns of Haddam and East-Haddam* (Middletown, 1814), 4. Both are plagued with factual errors, lack citations, and display strong ethnocentric biases, but they contain a few details not available from the other sources.

38. "Roger Williams to John Winthrop," *CMHS,* 6: 229; Stephen Hosmore to Thomas Prince, August 13, 1729, *CCHS* 3 (1890): 280–81; Field, *Statistical Account of Middlesex,* 62–63.

39. Stephen Hosmore to Thomas Prince, August 13, 1729, in *CCHS* 3 (1890): 280–81.

40. John Bishop to Increase Mather, August 11, 1678, "Mather Papers," *CMHS,* 4th Series 8 (1868): 306.

41. For the Montauk word "machee," see Silas Wood, *A Sketch of the First Settlement of the Towns*

on Long Island (Brooklyn, 1824), 28n. William Strachey, writing in the early seventeenth century, noted that Native Americans in the Chesapeake area used "mach" to mean "bad, enemy, evil." Philip Barbour notes that the northern Crees have a cognate, so probably all speakers of Algonquian languages shared some version of the term. See the compiled word lists and their attendant notes in Appendix 3 of Barbour, *Jamestown Voyages,* 2: 470. For the vertical cosmology of protective sounds above and malevolent sounds below, see Theresa Smith, "Island of the Anishnaabeg: An Interpretation of the Relationship Between the Thunder and Underwater Manitouk in the Traditional Ojibwe Life-World" (Ph.D. diss., Boston University, 1990).

42. "Roger Williams to John Winthrop," *CMHS,* 6: 229.

43. Hosmore to Prince, *CCHS* 3 (1890): 280–81.

44. According to R. B. Le Page and Andrée Tabouret-Keller, people perform acts of identity by projecting themselves onto a group or groups and then adjusting their own identities to better match those of the groups with which they identify. Individuals enact these group identities in their speech and social behaviors. A flaw in Le Page and Tabouret-Keller's model is that it holds the group to be unaffected by the actions of the individuals that constitute it. This caused problems with the statistical parts of their work, based on cluster analysis of Caribbean creole speech patterns. Allowing for the fact that the relation between groups and the individuals that compose them is reciprocal rather than unidirectional ruins "acts of identity" as a statistical psycho-social index of group identification, which was the authors' goal. However, this correction leaves their model, as they applied it, as a particularly useful description of the process of attributing identity to people, beings, or things outside oneself. I am putting Native Americans in the position of the researchers and putting the sources of the sounds in the position that Le Page and Tabouret-Keller assign to creole speakers. R. B. Le Page and Andrée Tabouret-Keller, *Acts of Identity: Creole-Based Approaches to Language and Ethnicity* (Cambridge, 1985).

45. Åke Hultkrantz, *The Religions of the American Indians,* trans. Monica Setterwall (Berkeley, 1979), 52; Myron Eels, "The Thunder Bird," *American Anthropologist,* old series 2 (1889): 335. For the geographical scope of thunder-being beliefs, see Hultkrantz, *Religions of the American Indians,* 50–52, 195n, 244; A. F. Chamberlain, "The Thunderbird amongst the Algonkins," *American Anthropologist,* old series 3 (1890): 51–54; William Healey Dall, "Masks, Labrets, and Certain Aboriginal Customs, with an Inquiry into the Bearing of their Geographical Distribution," *Annual Report of the Bureau of Ethnology* 3 (1882–83): 119–20; Smith, *Island of the Anishnaabeg,* 1, 17, 25, 140.

46. Le Jeune's "Relation of 1634," in *The Jesuit Relations and Allied Documents: Travels and Explorations of the Jesuit Missionaries in North America, 1610–1791,* ed. Edna Kenton (New York, 1954), 54.

47. Smith, "Island of the Anishnaabeg," 146, 144–45; Jack Frederick Kilpatrick and Anna Gritts Kilpatrick, ed., *Friends of Thunder, Folktales of the Oklahoma Cherokees* (Dallas, [1964]), 51; Margot Edmonds and Ella Elizabeth Clark, *Voices of the Winds: Native American Legends* (New York, 1989), 287. I have used the ethnohistorical technique of upstreaming here and below because I have been able to find documentary evidence from the seventeenth century that corroborates the modern ethnographies. This historically "anchored" upstreaming (or, as Axtell calls it, "downstreaming") merits more confidence than "inductive" upstreaming, which constructs a possible past based solely on comparative ethnographies. See James Axtell, *The European and the Indian: Essays in the Ethnohistory of Colonial North America* (Oxford, 1982), 9–10.

48. *H5N,* 153; *DRCHNY,* 4:987.

49. Smith, "Island of the Anishnaabeg," 87, 130–32, 154–55; Charles G. Leland, *Algonquin Legends* (New York, 1884) 263 n. 1.

50. *JR,* 12: 25–27.

51. *JR,* 10: 43–45.

52. Leland, *Algonquin Legends,* 259–67; Smith, "Island of the Anishnaabeg," 168–71.

53. *JR*, 10: 195–97; Smith, "Island of the Anishnaabeg," 178–79; John Bierhorst, *The Mythology of North America* (New York, 1985), 60–62, 90.

54. Roger Williams, *A Key into the Language of America* (London, 1643), 190–91, cited in Neal Salisbury, *Manitou and Providence: Indians, Europeans, and the Making of New England, 1500–1643* (New York, 1984), 37 (see also 37–39); Hultkrantz, *Religions of the American Indians*, 11–14; and Smith, "Island of the Anishnaabeg," vi.

55. *JR*, 5:57, 6: 157, 225, 10: 43–45, 195–97; Bruce G. Trigger, *The Huron: Farmers of the North* (Fort Worth, 1990), 107.

56. Smith, "Island of the Anishnaabeg," 157–58, 163. On thunders as the songbirds' "elders," see *JR*, 6: 160–61.

57. Leland, *Algonquin Legends*, 255–67.

58. *JR*, 10: 195–97, 12: 7–23.

59. *JR*, 58: 278–79. Thwaite's translation substitutes "lightnings" for the "thunders" (*tonnere*) reported in the original French (p. 279).

60. *JR* 57: 275; Smith, "Island of the Anishnaabeg," 166–67; Jeremiah Curtin and J. N. B. Hewitt, "Seneca Myths and Legends," *Annual Report of the Bureau of Ethnology* 32 (1910–11): 197–99, 229, 372; Erminnie A. Smith, "Myths of the Iroquois," *Annual Report of the Bureau of Ethnology* 2 (1880–81): 57–58; Leland, *Algonquin Legends*, 263–67.

61. *JR*, 10: 195–97, 6: 225, 12: 27; Smith, "Island of the Anishnaabeg," vi, 1–2, 17, 25.

62. Strachey, *Historie of Travell into Virginia Britania*, 99–100. For a similar but somewhat later and transformed version of this belief among Virginia Algonquian-speaking Indians that reinforces my reading of Strachey's purposely cryptic remarks, see Beverly, *History and Present State of Virginia*, 201.

63. Kenton, *Selections from the Jesuit Relations*, 54; *JR* 6: 173.

64. Jacques Marquette, *Of the First Voyage Made by Father Marquette Toward New Mexico, and How the Idea Thereof Was Conceived* (*JR* 5: 86–163), in Kenton, *Selections from the Jesuit Relations*, 338; Hennepin, *Description of Louisiana*, 72, 378, 381.

65. Hennepin, *Description of Louisiana*, 72; Peter Kalm, "Letter from Peter Kalm," in *A Journey from Pennsylvania to Onondaga in 1743 by John Bartram, Lewis Evans, and Conrad Weiser*, ed. Whitfield J. Bell, Jr. (Barre, Mass., 1973), 67. Also see Jonathan Carver, "Journals of the Travels of Jonathan Carver," in *The Journals of Jonathan Carver and Related Documents, 1766–1770*, ed. John Parker (St. Paul, 1976), 61 for Niagara; and Jonathan Carver, *Travels through the Interior Parts of North-America, in the Years 1766, 1777, and 1778* (London, 1778), 66, errata after 543 for St. Anthony's Falls.

66. Smith, "Myths of the Iroquois," 54–55.

67. Daniel K. Richter, *The Ordeal of the Longhouse: The Peoples of the Iroquois League in the Era of European Colonization* (Chapel Hill, N. C., 1992); and the following essays in Bruce G. Trigger, ed., *Handbook of North American Indians* (Washington, D.C., 1978): Thomas S. Abler and Elizabeth Tooker, "Seneca," 505–17; Elizabeth Tooker, "The League of the Iroquois: Its History, Politics, and Ritual," 418–41; Marian E. White, "Neutral and Wenro," 407–11; and Bruce G. Trigger, "Early Iroquoian Contacts with Europeans," 344–56.

68. Hennepin, *Description of Louisiana*, 381.

69. James Albert Ukawsaw Gronniosaw, *A Narrative of the Most Remarkable Particulars in the Life of James Albert Ukawsaw Gronniosaw, an African Prince, Written by Himself* (Newport, R.I., 1774), 9–10.

70. William D. Piersen, *Black Yankees: The Development of an Afro-American Subculture in Eighteenth-Century New England* (Amherst, Mass., 1988), 79, 205 n. 24; Cotton Mather, *Magnalia Christi Americana* (Cambridge, Mass., 1977), 2: 362.

71. Wyatt MacGaffey, *Religion and Society in Central Africa: The BaKongo of Lower Zaire* (Chicago, 1986), 204. On Shango, see Robert Farris Thompson, *Flash of the Spirit: African and Afro-American Art and Philosophy* (New York, 1983), 17, 84–95, 214.

72. Daniel Horsmanden, *The New York Conspiracy* (Boston, 1971), 297, cited in Piersen, *Black Yankees,* 79.

73. Piersen, *Black Yankees,* 79. On "Election Day," see Shane White, " 'It Was a Proud Day': African Americans, Festivals, and Parades in the North, 1741–1834," *Journal of American History* 81, no. 1 (1994): 13–50; Roger D. Abrahams, *Singing the Master: The Emergence of African American Culture in the Plantation South* (New York, 1992); Melvin Wade, "Shining in Borrowed Plumage: Affirmation of Community in the Black Coronation Festivals of New England, ca. 1750–1850," in *Material Life in America, 1600–1860,* ed. Robert Blair St. George (Boston, 1987), 171–82.

74. Peter Herndon, *Family Life among the Ashanti of West Africa,* Curriculum Unit 91.02.04 (New Haven, 1991), <http://www.yale.edu/ynhti/curriculum/units/1991/2/91.02.04.x.html>; K. Budu-Acquah, *Ghana, the Morning After* (London, 1975), 27–28. For rituals of possession, see Patricia Seed, "Taking Possession and Reading Texts: Establishing the Authority of Overseas Empires," *William and Mary Quarterly* 49, no. 2 (1992): 183–209, and Chapter 2 below.

75. Johannes Fabian maps out a version of this process in his *Time and the Other: How Anthropology Makes Its Object* (New York, 1983).

Chapter 2. From the Sounds of Things

1. For the approach to material culture used here, see Richard L. Bushman, *The Refinement of America: Persons, Houses, Cities* (New York, 1992); W. David Kingery, *Learning from Things: Method and Theory of Material Culture Studies* (Washington, D.C., 1996); Steven D. Lubar and W. D. Kingery, eds., *History from Things: Essays on Material Culture* (Washington, D.C., 1993); Robert Blair St. George, ed., *Material Life in America, 1600–1860* (Boston, 1988); and John Michael Vlach, *By the Work of Their Hands: Studies in Afro-American Folklife* (Charlottesville, 1991).

2. Michel Foucault, *The Order of Things: An Archaeology of the Human Sciences* (New York, 1973), xxii, 17–29. Many of the features of the enlightenment that Foucault describes were noticed independently and earlier by Marshall McLuhan, who attributed the shift toward atomization and categorization to be related to the introduction and dissemination of movable type. See Marshall McLuhan, *The Gutenberg Galaxy: The Making of Typographic Man* (New York, 1962), and Marshall McLuhan, *Understanding Media: the Extensions of Man* (New York, 1964). This section is concerned mostly with the discourse preceding the Enlightenment, however, which Foucault designates as the discourse of "*similitude.*"

3. A number of scholars find that the "Great Divide" theories of literacy and orality do not hold in present-day ethnographic situations that historians commonly project upon the oral past. Some, like Carol Fleischer Feldman, reject the divide altogether, while others consider the differences between literacy and orality to fall along a continuum. See Daniel Chandler, "Biases of the Ear and Eye: 'Great Divide' Theories, Phonocentrism, Graphocentrism & Logocentrism," *Media and Communications Site* (1995), http://www.aber.ac.uk/media/Documents/litoral/litoral.html; Carol Fleischer Feldman, "Oral Metalanguage," in *Orality and Literacy,* ed. David R. Olson and Nancy Torrance (Cambridge, 1991), 47–65. On face-to-face communication and orality, see Walter J. Ong, *Orality and Literacy: The Technologizing of the Word* (London, 1982); Walter J. Ong, *The Presence of the Word: Some Prolegomena for Cultural and Religious History* (New York, 1970); and Erving Goffman, "On Face Work," *Psychiatry* 18 (1955): 213–31. For extensions of the senses as markers of civilization, see McLuhan, *Understanding Media.* He adapts the idea from Edward T. Hall, *The Silent Language* (Greenwich, Conn., 1959).

4. Simon Harward, *A Discourse of Lightnings* (New York, 1971), sig. A1r, B1v–B2r. For bell beliefs as a kind of magic, see Keith Thomas, *Religion and the Decline of Magic: Studies in Popu-*

lar Beliefs in Sixteenth and Seventeenth-Century England (New York, 1971), 31–32, 49, 52–53, 57–58, 73–74.

5. Harward, *Discourse of Lightnings*, B2r. The priest would actually say the benediction in Latin (the translation is Harward's):

> *Omnipotens semiterne Deus tu hoc tintinnabulum coelestis benedictione persunde ut ante senitum eius longius effugentur ignita iacula inimici, percusio fulminum, impetus lapidum, læsio tempestatum.*

The italics in both the Latin and the translation are Harward's.

6. Thomas, *Religion and the Decline of Magic*, 31–33, 42, 49, 52–53, 57–58, 73–74.

7. Hayward, *Discourse of Lightnings*, sig. C3r.

8. David Cressy, *Bonfires and Bells: National Memory and the Protestant Calendar in Elizabethan and Stuart England* (Berkeley, 1989), 67, 70–76; and David Waldstreicher, *In the Midst of Perpetual Fetes: The Making of American Nationalism, 1776–1820* (Chapel Hill, N.C., 1997).

9. Cressy, *Bonfires and Bells*, 74. For English curfew bells and other bell customs see J. J. Raven, *Bells of England*, 317. For Rogation Day bell ringing see Thomas, *Religion and the Decline of Magic*, 62–65.

10. Cressy, *Bonfires and Bells*, 68–69.

11. The hogs were probably left by the Spanish or Portuguese years earlier. See Alfred W. Crosby, *The Columbian Exchange: Biological and Cultural Consequences of 1492* (Westport, Conn., 1977), 77–79. For a detailed description of the Bermudian food chain, see Strachey, *True Reportory*, 22–34. Sound played a role in keeping the castaways fed. Strachey reports that they caught great numbers of seabirds on the islands by "lowbelling," an old English technique in which the hunters went out at night with torches and used a bell to drive confused birds from their roosts into the hunters' nets. Strachey claimed that in an hour they once caught three hundred "cahows" this way (30–31). The cahow was named for its "strange hollo and harsh howling." Later, the accidental Bermudians discovered a daylight means of capturing the birds with vocal sounds instead of bells,

> which was by standing on the rocks or sands by the seaside and holloing, laughing, and making the strangest outcry that they possibly could. With the noise thereof the birds would come flocking to that place and settle on the very arms and head of him that so cried, and still creep nearer and nearer, answering the noise themselves; by which our men would weigh them by the hand, and which weighed heaviest they took for the best and let the others alone.

Cahows are now extinct. See Strachey, *True Reportory*, 31.

12. Strachey, *True Reportory*, 53.

13. For telegraph and radio as the beginning of disembodied communication, see Daniel J. Czitrom, *Media and the American Mind: From Morse to McLuhan* (Chapel Hill, N.C., 1982).

14. Strachey, *True Reportory*, 43–44, 45. On the power of spoken apologies or "unsaying," see Jane Neill Kamensky, *Governing the Tongue: The Politics of Speech in Early New England* (Oxford, 1998), 127–49.

15. Strachey, *True Reportory*, 63, 96.

16. *OED Online*, s.v. "earshot."

17. For a comparative discussion of these rituals of possession that does not take sound into account, see Patricia Seed, "Taking Possession and Reading Texts: Establishing the Authority of Overseas Empires," *William and Mary Quarterly*, 3rd series, 49, no. 2 (April 1992): 183–209. Strachey, *True Reportory*, 76, 78, 80; Smith, *Accidence*, 2: 23, 26–27; Smith, *Sea Grammar*, 2: 83, 102–3, 126–27; Percy, "Observations," 1: 144.

18. Father Andrew White, "A Brief Relation of the Voyage unto Maryland," in *Narratives of Early Maryland 1633–1684*, ed. Clayton C. Hall (New York, 1910), 30.

19. For more evidence and discussion of these claims, see Rath, "Early American Soundways." For "will" as an effective force in seventeenth-century Anglo-American culture, see Richard Cullen Rath, "'What Meanes Hee May for to Gett Her Over': The Transference of

Language and Culture from Old to New England in the Seventeenth Century" (honors thesis, Millersville University, 1991), 72–73.

20. White, "Voyage unto Maryland," 41; Smith, *Generall Historie,* in *CWCJS,* 2: 122, 170, 184; Roger Williams, *A Key into the Language of America* (London, 1643), 190–91, cited in Neal Salisbury, *Manitou and Providence: Indians, Europeans, and the Making of New England, 1500–1643* (New York, 1984), 37. See also Salisbury, *Manitou and Providence,* 37–39. The terms *Okeus* and *Manitec,* corresponding to the Iroquoian *Oki* and Algonkian *Manitou,* appear in the writings of both Smith and Strachey, indicating that the correspondence was more than a chance resemblance. For "rockets" as the precursor to "racket," see *OED Online,* s.v. "racket."

21. Smith, *True Relation,* in *CWCJS,* 1: 31, 37; Smith, *Proceedings,* in *CWCJS,* 1: 227 (another version appears in *Generall Historie,* in *CWCJS,* 2: 166).

22. Smith, *True Relation,* in *CWCJS,* 1: 41.

23. Strachey, *Historie of Travell into Virginia,* in *Jamestown Voyages,* 74; Smith, *Generall Historie,* in *CWCJS,* 2: 151 (also compared with others in 1: 14); Smith, *True Relation,* in *CWCJS,* 1: 73. For bravery, see *OED,* 2nd ed., s.v. "brave," "bravely," and "bravery." I first became aware of this meaning in Karen Ordahl Kupperman, *Indians and English: Facing Off in Early America* (Ithaca, 2000).

24. Strachey, *Historie of Travell into Virginia,* in *Jamestown Voyages,* 78; Henry Chandlee Forman, *Virginia Architecture in the Seventeenth Century* (Williamsburg, 1957), 8, 13, 14.

25. Smith, "Fragment G [1623?]" in *CWCJS,* 3:325 (insertion made by editor); William Strachey, *The Historie of Travell into Virginia Britania (1612)* (London, 1953), 105; Smith, *True Relation,* in *CWCJS,* 1: 57, 69; Smith, *Generall Historie,* in *CWCJS,* 2: 245.

26. Smith, *True Relation,* in *CWCJS,* 1: 45, 83.

27. Smith, *Generall History,* in *CWCJS,* 2: 178.

28. Strachey, *Historie of Travell into Virginia,* 59; Smith, *Generall History,* in *CWCJS,* 2: 169.

29. Smith, *Generall History,* in *CWCJS,* 2: 293, 296–97, 302–3.

30. Dell Upton, *Holy Things and Profane: Anglican Parish Churches in Colonial Virginia* (Cambridge, Mass., 1986), 250–51 n. 73.

31. Peter Benes, *New England Meeting House and Church, 1630–1850* (Boston, 1980), 59–62.

32. Elbridge H. Goss, "Early Bells of Massachusetts," *Proceedings of the Massachusetts Historical Society* (April and July 1874): 176–85, 279–88. Goss's article is a catalogue of references to bells, drums, and shells in seventeenth-century Massachusetts town records. Rather than referring to the individual town records, I have pointed the interested reader to Goss so that all the references rather than just the single reference will be available. The arguments about the significance of the bells and how they were used are my own.

33. Cressy, *Bonfires and Bells,* 203; Goss, "Early Bells of Massachusetts," 177, 180, 283–86; Huntington Historical Society, *Mustering and Parading: Two Hundred Years of Militia on Long Island, 1653–1868* (Huntington, N.Y., 1982), 8.

34. Goss, "Early Bells of Massachusetts," 177–78, 181, 282, 286; L. Elsinore Springer, *That Vanishing Sound* (New York, 1976), 20.

35. Edward Burrough, *A Declaration of the Sad and Great Martyrdom of the People of God Called Quakers in New England, for the Worshipping of God* (London, 1660), 24. Also see Carla Gardina Pestana, "The Quaker Executions as Myth and History," *Journal of American History* 80, no. 2 (1993): 441–69; and Cristine M. Levenduski, *Peculiar Power: A Quaker Woman Preacher in Eighteenth-Century America* (Washington, D.C., 1996), 20.

36. Edward Stickney and Evelyn Stickney, *Revere Bells* (Bedford, Mass., 1976), 4; Mather, *Illustrious Providences,* 170–71, 262–64.

37. Barbara Lambert and M. Sue Ladr, "Civic Anouncements: The Role of Drums, Criers and Bells in the Colonies," in *Music in Colonial Massachusetts, 1630–1820,* ed. Barbara Lambert (Boston, 1985), 869–933; Goss, "Early Bells of Massachusetts," 178; and Springer, *That Vanishing Sound,* 21. For sonic publishing and manuscript publishing as the norm in the

Chesapeake, see David D. Hall, "The Chesapeake in the Seventeenth Century," in *Cultures of Print: Essays in the History of the Book*, ed. David D. Hall (Amherst, Mass., 1996), 100–101.

38. Richard Olson, "Spirits, Witches, and Science: Why the Rise of Science Encouraged Belief in the Supernatural in Seventeenth-Century England," *Skeptic* 1, no. 4 (1992): 34–43; Robert Blair St. George, *Conversing by Signs: Poetics of Implication in Colonial New England Culture* (Chapel Hill, N.C., 1998), 157, 187; Mather, *Illustrious Providences*, 142; Cotton Mather, *Wonders of the Invisible World* (Boston, 1693), 65.

39. Goss, "Early Bells of Massachusetts," 280; Benes, *New England Meeting House and Church,* 59–62.

40. Springer, *That Vanishing Sound*, 18–19; *New England Weekly Journal*, Aug. 3, 1736: 21, 22 (487); *Boston Evening Post*, May 14, 1759: 31 (1237); Goss, "Early Bells of Massachusetts," 280; and George Keith, *The Presbyterian and Independent Visible Churches in New England and Elsewhere Brought to the Test* (London, 1691), 166.

41. Goss, "Early Bells of Massachusetts," 280–82; and Benes, *New England Meeting House and Church,* 59–62; Marian C. Donnelly, *The New England Meeting Houses of the Seventeenth Century* (Middletown, Conn., 1968), 45, 66.

42. Goss, "Early Bells of Massachusetts," 177–80, 281–83.

43. This section is adapted from Richard Cullen Rath, "African Music in Seventeenth-Century Jamaica: Cultural Transit and Transition," *William and Mary Quarterly* 50, no. 3 (1993): 700–726. The technical details of the arguments presented below will be found there.

44. William Lux, *Historical Dictionary of the British Caribbean* (Metuchen, N.J., 1975), 120, 135, 141, 157, 158. The imaginative reconstruction that follows is, unless otherwise noted, based on Sir Hans Sloane, M.D., *A Voyage to the Islands of Madera, Barbados, Nieves, St. Christopher and Jamaica . . .* (London, 1707), 1: xlviii, xlix, 2: plate 232.

45. Sloane, *Voyage to the Islands,* 1: xlviii, xlix. For the significance of sweeping a clearing, see Wyatt MacGaffey, *Religion and Society in Central Africa: The BaKongo of Lower Zaire* (Chicago, 1986), 51, 54–56, 127–31; Robert Farris Thompson, "Kongo Influences on African-American Artistic Culture," in *Africanisms in American Culture*, ed. Joseph Holloway (Bloomington, Ind., 1990), 164; for the use and significance of rings of people in African American music, see Sterling Stuckey, *Slave Culture: Nationalist Theory and the Foundations of Black America* (New York, 1987), 1–90; for the use of an upturned clay pot or hollowed-out gourd as a functional substitute for the skin drum, see J. H. Kwabena Nketia, *The Music of Africa* (New York, 1974), 74.

46. E. Franklin Frazier, *The Negro Family in the United States*, rev. ed. (Chicago, 1948), 1–69; Melville J. Herskovits, *Myth of the Negro Past* (Boston, 1941); Orlando Patterson, *The Sociology of Slavery: An Analysis of the Origins, Development and Structure of Negro Slave Society in Jamaica* (Cranbury, N.J., 1967), 10, 74, 80–86, 151, 178, 284; Edward Brathwaite, *The Development of Creole Society in Jamaica, 1770–1820* (Oxford, 1971), 96–101, 212–39, 298, 306–11; Sidney W. Mintz and Richard Price, *An Anthropological Approach to the Afro-American Past: A Caribbean Perspective*, ISHI Occasional Papers in Social Change, No. 2 (Philadelphia, 1976), 5–7, 21, 24–27, 46, *passim*. A revised edition with a new preface has been published under the title *The Birth of African-American Culture: An Anthropological Perspective* (Boston, 1992); Barbara Kopytoff, "The Development of Jamaican Maroon Ethnicity," *Caribbean Quarterly* 22 (1976): 33–50; Winifred Vass, *The Bantu Speaking Heritage of the United States* (Los Angeles, 1979); Patricia Jones-Jackson, *When Roots Die: Endangered Traditions on the Sea Islands* (Athens, Ga., 1987); John Thornton, *Africa and Africans in the Making of the Atlantic World, 1400–1680* (Cambridge, 1992), 183–234; Daniel C. Littlefield, *Rice and Slaves: Ethnicity and the Slave Trade in Colonial South Carolina* (Baton Rouge, 1981); Charles Joyner, *Down by the Riverside: A South Carolina Slave Community* (Chicago, 1984); Joseph Holloway, ed., *Africanisms in American Culture*, particularly the essays by Holloway, Robert L. Hall, Jason Gaston Mulira, Robert Farris Thompson, and Margaret Washington Creel; Gwendolyn Midlo Hall, *Africans in Colonial Louisiana: The Development of Afro-Creole Culture in the Eighteenth Century* (Baton

Rouge, 1992); Ira Berlin, *Many Thousands Gone: The First Two Centuries of Slavery in North America* (Cambridge, Mass., 1998); Michael Angelo Gomez, *Exchanging Our Country Marks: The Transformation of African Identities in the Colonial and Antebellum South* (Chapel Hill, 1998); Philip D. Morgan, *Slave Counterpoint: Black Culture in the Eighteenth-Century Chesapeake and Lowcountry* (Chapel Hill, 1998).

47. Kenneth Bilby, "The Caribbean as a Musical Region," in *Caribbean Contours,* ed. Sidney W. Mintz and Sally Price (Baltimore, 1985), 186.

48. I am not claiming any special objectivity on Baptiste's part, nor am I contending that his transcriptions are more accurate (or even nearly as accurate) as those of modern ethnomusicologists. Western musical notation is unable to capture many features of non-European musics, and I have no evidence that Baptiste's recording transcended any of these limits. However, the interpretation below accounts for the many non-European features that are captured in his transcriptions, features also found in the modern ethnomusicological studies of West Africa, Central Africa, and Jamaica.

49. For definitions of pidginization, see John A. Holm, *Pidgins and Creoles* (Cambridge, 1988), 1: 3–70; Derek Bickerton, *Language and Species* (Chicago, 1990), 105–29; Derek Bickerton, "Creole Languages," *Scientific American* 249 (1983): 116–21; Derek Bickerton, *Roots of Language* (Ann Arbor, 1981), 1–42, esp. 1–8; Ronald Wardaugh, *An Introduction to Sociolinguistics* (Oxford, 1986), 70–76; Barbara Lalla and Jean D'Costa, *Language in Exile: Three Hundred Years of Jamaican Creole* (Tuscaloosa, 1990), 1–6.

50. For planters' structural power see Patterson, *Sociology of Slavery,* 80, 86, 178. For structure/content division, and cultural isolation of both planters and enslaved, see Brathwaite, *Development of Creole Society,* 96–101, 212–39, 306–9. For the concept of ethnic mixing as a strategy adopted by planters see Mintz and Price, *An Anthropological Approach to the Afro-American Past,* 8–10. Some planters preferred particular ethnicities, especially Akans in Jamaica, a factor that tempers Mintz and Price's hypothesis. Evidence from shipping lists further qualifies the mixing hypothesis, because even if planters adopted that strategy, supplies of ethnically homogenous slave cargoes often precluded meeting such demands (Thornton, *Africa and Africans in the Making of the Atlantic World,* 192–205).

51. J. H. Kwabena Nketia, "The Musical Languages of Subsaharan Africa," *African Music: Proceedings from a Meeting in Yauondé, Cameroon, 23–27 February 1970* (Yauondé, 1972), 7–42.

52. Patterson, *Sociology of Slavery,* 134–36; Philip D. Curtin, *The Atlantic Slave Trade: A Census* (Madison, Wisc., 1969), 56–62, 138, 150, 158–62; John Storm Roberts, *Black Music of Two Worlds* (New York, 1972), 38; Patterson, "Slavery and Slave Revolts," in *Maroon Societies,* ed. Richard Price (Garden City, N.Y., 1973), 256, 283. For information on manpower, tactics, and battles in this region of Africa see Ray Kea, *Settlement, Trade and Polities on the Seventeenth Century Gold Coast* (Baltimore, 1982), 130–68 (esp. 142 for a sampling of battles and their casualties, and 159 for the role of the English, who had a fort at Kormantse). Other modern Jamaican Maroon communities exist in the parishes of St. Mary, Portland, and St. Thomas. None exist today in St. Ann or Clarendon (Kenneth Bilby, personal communication, January 1992).

53. Sloane *Voyage,* xlix. For the intellectual origins of Central African attitudes toward enslavement to Europeans see Wyatt MacGaffey, *Religion and Society in Central Africa: The Bakongo of Lower Zaire* (Chicago, 1986), 62. For ethnic patterns of slave imports, see Patterson, *Sociology of Slavery,* 134–36; Curtin, *Atlantic Slave Trade,* 56–62, 138, 150, 158–62. Curtin arrives at a "speculative" 6.3 percent Gold Coast share of total slave imports to Jamaica during the period 1685–92 by subdividing the undifferentiated "Guinea" region of his source for Jamaica by the differentiated (Guinea was considered to comprise Sierra Leone, the Windward Coast, and the Gold Coast) ethnic distribution for the English slave trade as a whole during this time. He qualifies his findings by noting, first, that in contemporary accounts planters expressed a preference for Gold Coast slaves in spite of their bellicosity; and second, that Ja-

maica planters purchased about 80 percent of all Gold Coast slaves sold in the colonies until about 1790. Even after assuming that, as a result of Koromanti-led revolts, Gold Coast imports to Jamaica were at their lowest precisely at the time documented by Curtin's source. Patterson's higher Gold Coast figures, which also account for the influx of Angolans, make more sense for the late seventeenth century.

54. For locations of, and information on, the seventeenth-century slave trade on the Bight of Benin, particularly Popo, see Robin Law, *The Slave Coast of West Africa, 1550–1750: The Impact of the Atlantic Slave Trade on an African Society* (Oxford, 1991), 5, 9, 14–19, 129–35, and Thornton, *Africa and Africans in the Making of the Atlantic World*, xii, xxiv.

55. Drouin De Bercy, *De Saint-Domingue* (Paris, 1814), cited in Pierre Pluchon, *Vaudou, Sorciers, Empoisoneurs De Saint-Domingue ç Haiti* (Paris, 1987), 113; Jean Baptiste du Tertre, *Histoire Générale des Antilles Habiteés par les François* (Paris, 1667), 2: 526–27; Richard Ligon, *A True and Exact History of the Island of Barbados* (London, 1657), 46–49, 50, 52, 107; John Gabriel Stedman, *Narrative of a Five Years Expedition against the Revolted Slaves of Surinam* (from the 1790 manuscript), ed. Richard Price and Sally Price (Baltimore, 1988), 516, 538–41. The most complete collection of early American African–derived musical sources is Dena Epstein, *Sinful Tunes and Spirituals: Black Folk Music to the Civil War* (Chicago: 1977), 1–46 and elsewhere. She points out that mainland North American sources are unknown until the eighteenth century (xvi).

56. Sloane, *Voyage*, xlvii–lii.

57. Nketia, "Musical Languages," 19–31; Nketia, *African Music in Ghana*, 34–61.

58. A. M. Jones, *Studies in African Music* (London, 1959) 1: 224–31, and Rose Brandel, *The Music of Central Africa* (The Hague, 1962), 60–62.

59. Nketia, *African Music in Ghana*, 43–59.

60. Ibid., 64.

61. Nketia, *The Music of Africa*; A. M. Jones, *Studies in African Music*; J. S. Roberts, *Black Music of Two Worlds*, 20–23; Kenneth Bilby, "The Kromanti Dance of the Windward Maroons of Jamaica," *Nieuwe West-Indische Gids* 55 (1981): 71–76.

62. Alan P. Merriam, "African Musical Rhythm and Concepts of Time Reckoning," in *African Music in Perspective* (London, 1982), 443–62; Ruth M. Stone, "In Search of Time in African Music," *Music Theory Spectrum* 7 (1981): 139–48; David Reck, *Music of the Whole Earth* (New York, 1977), 164–65, 180.

63. Julius Scott, "The Common Wind: Currents of Afro-American Communication in the Era of the Haitian Revolution" (Ph.D. diss., Duke University, 1986); Peter Linebaugh and Marcus Buford Rediker, *The Many-Headed Hydra: Sailors, Slaves, Commoners, and the Hidden History of the Revolutionary Atlantic* (Boston, 2000).

64. Georgia Writers Project, *Drums and Shadows: Survival Studies among the Georgia Coastal Negroes* (1940; Athens, Ga., 1992), 180; Lydia Parrish, *Slave Songs of the Georgia Sea Islands* (1942; Athens, Ga., 1992). Johnson's words were written as "he use tuh beat duh drum tuh duh fewnul, but Mr Couper he stop dat. He say he dohn wahn drums beatin roun duh dead." I have paraphrased in order to defer issues of dialect misrepresentation in the WPA and Federal Writer's Project collections. See Edgar W. Schneider, *American Earlier Black English: Morphological and Syntactical Variables* (Tuscaloosa, 1989), 1–16, 42–53. Slaves' names often reflected their place of origin or where they were purchased. The *Ndembu* or *Dembo* area of Central Africa comprised a politically unstable periphery of the Kingdom of Kongo, making Dembo most likely a captive imported from that region. For *Ndembu*, see Thornton, *Africa & Africans in the Making of the Atlantic World, 1400–1680* (Cambridge, 1992), xxxi–xxxiii. For naming a slave after a region see Newbell Niles Puckett, "Names of American Negro Slaves," in *Motherwit from the Laughing Barrel: Readings in the Interpretation of Afro-American Folklore*, ed. Alan Dundes (New York, 1981), 159. For Couper, see Malcom Bell, Jr., *Major Butler's Legacy* (Athens, Ga., 1987), 398. Couper lived until 1866, managing the family's plantation. His sons

were both killed in the Civil War, and his father had died in 1850, before Johnson was born. Thus, Dembo was most likely drumming while he was enslaved.

65. Girolamo Merolla, *A Voyage to Congo and Several Other Countries, Chiefly in Southern-Africk, in a Collection of Voyages and Travels*, ed. Awnsham Churchill (1682; London, 1704), 1: 651– 756. For Songo, see Thornton, *Africa and Africans*, xxxv.

66. Nieman's letter appears in Adam Jones, ed., *German Sources for West African History 1599–1699* (Wiesbaden, 1983), 88. For von der Groeben's portrayal of the Gold Coast see Jones, *German Sources*, 52. Another description of Akan state instruments and ensembles appears in Pieter de Marees, *Description and Historical Account of the Gold Kingdom of Guinea (1602)* (Oxford, 1987), 93, 171–72.

67. J. H. Kwabena Nketia, "History and Organization of Music in West Africa," in *Essays on Music and History In Africa*, ed. K. P. Wachsmann (Evanston, Ill., 1971), 17–22.

68. Dena Epstein, *Sinful Tunes and Spirituals: Black Folk Music to the Civil War* (Urbana, 1977), 58–60, 62. For a general account of Antigua, see David Barry Gaspar, *Bondmen and Rebels: A Study of Master-Slave Relations in Antigua* (Baltimore, 1985). While drumming was not an issue at the actual inquest, the fact that it was reported so in the North American press underscores Anglo-American elites' fears in this regard. See *Pennsylvania Gazette*, March 10–17, 1737; March 17–24, 1737.

69. On the uses of drums for displays of state power, particularly in Kwa regions, see T. Edward Bowditch, "Excerpts from *Mission from Cape Coast Castle to Ashantee*," in *Readings in Black American Music*, ed. Eileen Southern (1819; New York, 1983), 9–15; J. H. Kwabena Nketia, "History and Organization of Music in West Africa," in *Essays On Music and History in Africa*, ed. Klaus Wachsmann (Evanston, 1971), 17–22; J. H. Kwabena Nketia, *African Music in Ghana* (Evanston, 1963), 47–48, 103; J. H. Kwabena Nketia, *The Music of Africa* (New York: W. W. Norton, 1974), 167–70; Meki Nzewi, "Traditional Strategies for Mass Communication: The Centrality of Igbo Music," in *Selected Reports in Ethnomusicology* (Berkeley: University of California Press, 1984), 5: 318–28; R. S. Rattray, *Akan-Ashanti Folk-Tales* (Oxford: Clarendon Press, 1930), 133–34. Kwa languages are spoken from the Ibo region of Eastern Nigeria to the westernmost area of present-day Ghana. According to glottochronologists, a single Kwa language began to fragment about 8,000 years ago, and now the Eastern and Western language branches are only marginally related. All of them are tonemic, however. At the center, Yoruba speakers use nine tones. To the west of this tonal "hearth," Western Kwa languages generally have two to four tones. See William E. Welmers, *African Language Structures* (Berkeley: University of California Press, 1973).

70. For Central African state drumming traditions that resembled Gold Coast practices in many surface features, see Michael Angelo and Denis de Carli, *A Curious and Exact Account of a Voyage to Congo in the Years 1666 and 1667*, trans. A. and J. Churchill, in *A Collection of Voyages and Travels* (London, 1704), 1: 622, and Jos Gansemans and Barbara Schmidt-Wenger, *Musikgeschichte in Bildern: Zentralafrika* (Leipzig, 1986), 15–27.

71. De Marees, *Historical Account of the Gold Kingdom*, 93, 171–72. Nieman's account is in Jones, ed., *German Sources*, 88.

72. Helen Roberts, "Possible Survivals of African Song in Jamaica," *Musical Quarterly*, 22 (1926): 346.

73. John Thornton, personal communication, March 1990; for modern illustrations see MacGaffey, *Religion and Society in Central Africa*, 125.

74. Nketia, *African Music in Ghana*, 31–32.

75. Angelo and Carli, *Voyage to Congo*, in Churchill, *Voyages and Travels*, 1: 622; Ligon, *History of Barbados*, 48–49. Another Capuchin missionary, Giovanni Antonio Cavazzi, also left accounts and pictures of Central African instruments in his *Istorica Descrizione de' tr' Regni Congo, Matamba, et Angola* (Bologna, 1687), 157, 211, 216. For reproductions of a number of privately held oil paintings by Cavazzi, see Ezio Bassanni, "Un Cappuccino nell'Africa nera del

Seicento," *Quaderni Poro* 4 (1987), and Gansemans and Schmidt-Wenger, *Musikgeschichte in Bildern: Zentralafrika*, 15–27. The latter also reproduces sketches by Merolla and others.

76. Stedman, *Narrative of a Five Years Expedition*, 438–40. For a historical discussion of Central African musical styles that includes both sansas and marimbas, see Gansemans and Schmidt-Wenger, *Musikgeschichte in Bildern: Zentralafrika,* 17–27.

77. Gerhard Kubik, *Westafrika: Musikgeschicte in Bildern* (Leipzig, 1989), 140–80.

78. See Thornton, *Africa and Africans,* 207–8, for the relative ease with which seventeenth-century varieties of Kikongo, Kimbundu, and Fon (an Eastern Kwa language) have been translated to their modern equivalents.

79. For idealization of African ethnic identities, see Kopytoff, "Maroon Ethnicity"; John Thornton, "War, the State, and Relgious Norms in 'Coromantee' Thought: The Ideology of an African American Nation," in *Possible Pasts: Becoming Colonial in Early America,* ed. Robert Blair St. George (Ithaca, 2000), 181–200; John Thornton, "The Coromantees: An African Cultural Group in Colonial North America and the Caribbean," *Journal of Caribbean History* 32/1-2 (1998): 161–78; Thornton, *Africa and Africans,* 220.

80. *Records in the British Public Records Office,* trans. 13, 196; *South Carolina Commons House Journals* 1702, 64–65 1707–08, 53; all cited in Peter Wood, *Black Majority: Negroes in Colonial South Carolina from 1670 through the Stono Rebellion* (New York, 1975), 125.

81. For the Charleston "conspiracy" see *Pennsylvania Gazette,* October 29–November 5, 1730; also *Boston Weekly Newsletter,* October 22, 1730 (cited in Wood, *Black Majority,* 299). The letter itself was dated August 20, 1730, in both papers.

82. *South Carolina Gazette,* May 19–May 26, 1733. Wood, *Black Majority,* 244–45 for remarks on Vander Dussen's (also spelt as Vanderdussen) temperament.

83. For *capoeira,* see John Storm Roberts, *Black Music of Two Worlds* (New York, 1972), 27–28; John Lowell Lewis, *Ring of Liberation: Deceptive Discourse in Brazilian Capoeira* (Chicago, 1992). For musical bows, see Gansemans and Schmidt-Wenger, *Musikgeschichte in Bildern,* 127–31; Gerhard Kubik, "Capoeira Angola and Berimbau," in *Angolan Traits in Black Music Games and Dances of Brazil: A Study of African Cultural Extensions Overseas* (Lisboa, 1976), 27–36.

84. John K. Thornton, "The Art of War in Angola," *Comparative Studies in Society and Culture* 30 (April 1988): 362–65.

85. For *kalinda* see Roberts, *Black Music of Two Worlds,* 26–27, 115–116, 123, 157; and Epstein, *Sinful Tunes and Spirituals,* 24, 28, 30–38, 82, 92, 94, 135. For batons see Thompson, "Kongo Influences on African-American Artistic Culture," 162–63, 182–83. For Cuba see Odilio Urfe, "Music and Dance In Cuba," in *Africa in Latin America,* ed. Manuel Moreno Fraginals (New York, 1984), 170–188, esp. 173, 176, 181, 183, 185. For "knocking and kicking" see G. Daniel Dawson's liner notes in Grupo de Capoeira Angola Pelourinho, *Capoeira Angola: Salvador, Brazil,* sound recording (Washington D.C., 1996). For *maculelê,* see Bira Almeida, *Capoeira, a Brazilian Art Form: History, Philosophy, and Practice* (Berkeley, 1986) 46–47 n. 8, 159.

86. "Extract of a Letter from South Carolina Dated October 2," *Gentleman's Magazine* 10 (1740): 127–29, cited in *American Negro Slavery: A Documentary History,* Michael Mullin, ed. (Columbia, S.C., 1976), 85; *South Carolina Commons House Journals, 1739–1741* (Columbia, S.C.: 1907–1946, 1951–1962), 84, cited in Wood, *Black Majority,* 321; "Account of the Negroe Insurrection in South Carolina" in *Colonial Records of the State of Georgia,* 26 vols., ed. Allan D. Chandler and Lucien L. Knights (Atlanta, 1904–16), 23, ii: 233, cited in Wood, *Black Majority,* 314–20.

87. Peter Wood refers to the twenty slaves who formed the core of the revolt as "Angolans." Wood, *Black Majority,* 314. John Thornton presents military, contextual, and religious evidence indicating that the core group was from the closely related Kingdom of Kongo. John K. Thornton, "African Dimensions of the Stono Rebellion," *American Historical Review* 96 (1991).

88. Wood, *Black Majority*; Thomas Cooper and David J. McCord, eds., *The Statues at Large: South Carolina* (Columbia, 1836–41), 8 (1840): 410; William A. Hotchkiss, ed., *A Codification of the Statute Law of Georgia, Including the English Statutes of Force* (Savannah, 1845), 813; all cited in Epstein, *Sinful Tunes and Spirituals*, 59, 60, 62.

89. Epstein, *Sinful Tunes and Spirituals*, 60.

90. For musicians as status symbols to planters and Charleston as a musical center see Raoul F. Camus, *Military Music of the American Revolution* (Chapel Hill, N.C., 1976), 54–55. For the demand for slave musicians see Tilford Brooks, *America's Black Musical Heritage* (Englewood Cliffs, N.J., 1984), 164–68. For "hiring out" see Epstein, *Sinful Tunes and Spirituals*, 80.

91. The earliest extant American violin manual is Francesco Geminiani, *An Abstract of Geminiani's Art of Playing on the Violin* (Boston, 1769), 1, 5–6, 10. The only extant copy is bound from three or four partial sets of pages that had been combined to make a single thirteen-page volume (thanks to Daniel Slive for interpreting the binding and paper). The John Carter Brown Library purchased that copy from a British bookseller. Not all the musical notation is printed, either, indicating that the Boston printers may have been unable to deliver the pamphlet to market at all. A 1763 advertisement in Newport, Rhode Island, announced "Crone's Rules to play the Fiddle Well, without a Master, by Way of Question and Answer, with 25 copper Plates." See Arthur R. LaBrew, *Black Musicians of the Colonial Era 1700–1800: Preliminary Index* (Detroit, 1977), 12. Both manuals seem to have been aimed at New England, rather than southern African American readers, though. On what slaves might be likely to read, see Samuel Davies, *The State of Religion in Virginia, Particularly among the Negroes,* 2nd ed. (London: R. Pardon, 1757).

92. Moreau de Saint-Méry, Médéric Louis âlie, *Déscription topographique, physique, civile, politique, and historique de la Partie Française de l'Isle Saint-Dominingue* (Philadelphia, 1797), 1: 51; the "rough translation" of the cited passage into English is made by Epstein, *Sinful Tunes and Spirituals*, 116.

93. For increased imports from Mende/Western Atlantic groups after Stono see Creel, *A Peculiar People*.

94. Henry William Ravenal, "Recollections of Southern Plantation Life," *Yale Review* 26 (June 1936): 768–69; also 750, 774–75. Also see Epstein, *Sinful Tunes and Spirituals*, 77–87, 114–17, 120–24.

95. The pattern of drummers holding a stick in one (usually the right) hand is found throughout West and Central Africa. It is most pronounced north of the Kongo/Angola region, where playing with two hands—or less often, with two sticks—predominates, but playing with either one hand or one stick is not uncommon either. For seventeenth-century visual evidence of this, see the reproductions in Gerhard Kubik, *Westafrika: Musikgeschicte in Bildern,* 47, 49, 67, 69, 73, 79, 84, 99, 113, 114; and Gansemans, *Zentralafrika,* 17, 19, 51, 62, 68, 71, 95, 99, 108, 123, 165–67. For an image of a Jamaican fiddler holding a fiddle as a drum with the bow as a stick, see Richard Cullen Rath, "Drums and Power: Ways of Creolizing Music in Coastal South Carolina and Georgia, 1730–1790," in *Creolization in the Americas: Cultural Adaptations to the New World,* ed. Steven Reinhardt and David Guisseret (Arlington, Tex., 2000). For stringless violin-shaped instruments played with a stick see Cavazzi, *Istorica, Descrizione de' tr' Regni Congo, Matamba et Angola,* 200. Many illustrations from both can be found in Gansemans and Schmidt-Wenger, *Zentralafrika,* 15–27, 127–31.

96. David C. Barrow, "A Georgia Corn Shucking," *Century Magazine* 24 (1882): 878, cited in Roger D. Abrahams, *Singing the Master: The Emergence of African American Culture in the Plantation South* (New York, 1992), 103 n. 40, 186; Nettie Powell, *A History of Marion County, Georgia* (Columbus, Ga., 1931), 33, cited in Abrahams, *Singing the Master,* 186; William C. Handy, *Father of the Blues* (New York, 1941), 5, cited in Abrahams, *Singing the Master,* 103 n. 40, 186. For the prevalence of related practices in Latin America and the Caribbean, see Isabel Aretz, "Music and Dance in Continental Latin America, with the Exception of Brazil," in *Africa in*

Latin America, 197. For "beating straws" in Anglo-American fiddling in Alabama, see Joyce H. Cauthen, *With Fiddle and Well-Rosined Bow: Old-Time Fiddling in Alabama* (Tuscaloosa, 1989).

97. Ravenal, "Recollections of Southern Plantation Life," 768–69. The ring and the shuffling step and their implications as pan-Africanisms are discussed at length in Stuckey, *Slave Culture: Nationalist Theory and the Foundations of Black America.* Thornton (personal communication, March 10, 1993) has found similar accounts of this behavior as depicted by Italian missionaries to Kongo in the 1690s. See Marcellino d'Atri, *L'anarchia congolese,* fol. 335 of the original manuscript (p. 158 of Carlo Tosso's edition) and Luca da Caltanisetta's manuscript, fol. 60 (p. 290 of Romain Rainero's edition). Neither of these accounts describes military displays, but the functional divide between forms of entertainment and forms of training may not have existed.

98. Compiled from LaBrew, *Black Musicians of the Colonial Period,* 115–22. The author lists two men, Prince Lewis and Ketto, among the Hessian recruits whose race was unknown. The names "Prince" and "Cato" were almost exclusively used among slaves, so these two were listed as being of African descent in the compiled totals.

99. For American military drumming see Camus, *Military Music of the American Revolution,* 8–9, 128–50; also Eileen Southern, *The Music of Black Americans: A History,* 2nd ed. (New York, 1983), 65.

100. Henry George Farmer, *The Rise and Development of Military Music* (1912; Freeport, New York, 1970), frontispiece, 66, 70–78; Camus, *Military Music of the American Revolution,* 35–39, 122; LaBrew, *Black Musicians of the Colonial Period,* in which appear between pages 122 and 123 three illustrations of members of Hessian regiments who served in the American war. All three depict black drummers. See also LaBrew, *Black Musicians of the Colonial Period,* 99–122.

101. Henry G. Farmer, "The Turkish Influence in Military Music," in *Handel's Kettledrums and Other Papers on Military Music* (London, 1950), 46; cited in Raoul F. Camus, "The Military Band in the United States Army prior to 1834" (Ph.D. diss., New York University, 1969), 130.

Chapter 3. No Corner for the Devil to Hide

1. *Encyclopedia Brittanica,* s.v. "Acoustics." The source document was torn from the original volume and is not labeled. The attribution to the 1787 edition is from the finding aid, the attribution to 1792 from the archivist; in Warshaw Collection, Acoustics box, folder 2, Archive Center, National Museum of American History, Smithsonian Institution, Washington, D.C. "Diacoustics" included the generation of voices as well as sounds of instruments. "Catacoustics" often concerned the manipulated reflection of voices as well as other sounds. Today this field is known as architectural acoustics.

2. I have reconstructed the acoustics of the chancel from descriptions found in G. W. O. Addleshaw and Frederick Etchells, *The Architectural Setting of Anglican Worship: An Inquiry into the Arrangements for Public Worship in the Church of England from the Reformation to the Present Day* (London, 1948), 15–22. The lute analogy is my own. For the tympanum, see *OED Online,* s.v. "tympanum."

3. Michael Forsyth, *Buildings for Music: The Architect, the Musician, and the Listener from the Seventeenth Century to the Present Day* (Cambridge, Mass., 1985), 4–13; Hope Bagenal, "Bach's Music and Church Acoustics," *Journal of the Royal Institute of British Architects* 37, no. 5 (1930): 154–63.

4. Stephen Palmer Dorsey, *Early English Churches in America, 1607–1807* (New York, 1952), 16; Addleshaw and Etchells, *Architectural Setting of Anglican Worship,* 54, 245–53; Marian C. Donnelly, *The New England Meeting Houses of the Seventeenth Century* (Middletown, Conn., 1968), 36.

5. In fact they are two separate signals, with the original louder than the reflection, but

if an echo occurs about one-thirtieth of a second or less after the original sound, the ear cannot distinguish them. The two together are perceived as a single "fat" signal that is louder than the unreflected one because it consists of the unreflected plus the reflected signal.

6. Forsyth, *Buildings for Music,* 9; Addleshaw and Etchells, *Architectural Setting of Anglican Worship,* 22–63, 68–86, 245–53. Dell Upton makes the case that the cover over the pulpit was an ornament indicating power, and that it was called a "type" or "canopy" rather than a tester or sounding board. I believe that both Upton's definition and a definition that treats them as acoustic devices can be sustained simultaneously. It is true that many testers, particularly in larger churches built at later dates in the eighteenth century, were ineffective both in their construction and placement as acoustic devices. It is equally true that many of them were built and placed in such a way as to direct the voice and amplify it. Where, how, and when testers were used acoustically is the focus of this study, not whether. The earliest reference to "sounding board" referring specifically to the apparatus over the pulpit in the *OED Online* is from 1766, which would place it in the colonial period. It is important to remember that vernacular usage of the term no doubt antedated the first published citation. References to musical instrument "sound boards" occur throughout the sixteenth and seventeenth centuries, including one referring to a musician's attempt to play the harmonic stops of a monochord in which "none wold speke" because "Þe sownd~borde was to hy." This shows that the speech and sound boards were at least conceptually related in the sixteenth century. The *OED Online* dates "testers" with the meaning of pulpit coverings to 1908. Prior to that the word referred to bed canopies or canopies placed over dignitaries, which would support Upton's usage if earlier references are found. The "tipe" or "type" is not listed in the *OED Online* as a pulpit cover in any of the references used to support the definition. Although the etymology of "tipe" as a canopy is regarded by the *OED Online* as unknown, the evidence presented points toward the same etymology as for other entries for "type," which share the same Greek root, τύπτειν, meaning "to strike or beat, as in a drum"), from which "tympanum" is derived. See Dell Upton, *Holy Things and Profane: Anglican Parish Churches in Colonial Virginia* (Cambridge, Mass., 1986), 134–35; and *OED Online,* s.v. "type," "tipe," "tympanum," "tester," "sound-board," and "sounding board."

7. For an excellent discussion of the Globe's acoustics, as well as the acoustics of other Elizabethan theaters, see Bruce R. Smith, *The Acoustic World of Early Modern England: Attending to the O-factor* (Chicago, 1999), 206–17. I disagree with Smith's formulation of "two liturgical ideas—one [Catholicism] based on vision, the other [Protestantism] on audition." The Reformation adopted soundways that valued characteristics of the sonic pallette different from those valued by Catholicism, as the Catholic emphasis on forcefulness and reverberation compared to the Protestant emphasis on clarity underscores. Arguments could be made for Protestantism to be more visual than Catholicism, in that a much greater proportion of the devout were expected to be literate enough to read the Bible than in Catholicism, where only the clergy were expected to be able to do so. Although vision was plainer for Protestants, with images removed as objects of learning, reading is still the taking in of language through the eyes. Again, the better argument would show *how* vision ways differed between the two branches of Christianity rather than asserting the primacy of one sense over another. The difference does not affect the main thrust of Smith's argument. See Smith, *Acoustic World of Early Modern England,* 261.

8. For the distinction between "auditory" and "Basilica" type churches, see Addleshaw and Etchells, *Architectural Setting of Anglican Worship,* 52–62, and Upton, *Holy Things and Profane,* 56–59.

9. William Strachey, *A True Reportory of the Wreck and Redemption of Sir Thomas Gates, Knight, upon and from the Islands of Bermudas,* in *A Voyage to Virginia in 1609,* ed. Louis B. Wright (Charlottesville, 1964), 80; and Dorsey, *Early English Churches in America,* 16–23.

10. Upton, *Holy Things and Profane,* 59.

11. Ibid., 59, 175–96.

12. Ibid., 59, 175–96. Jamestown's old brick church was forty-six feet at the tip of the ceiling. See James Scott Rawlings, *Virginia's Colonial Churches: An Architectural Guide; Together with Their Surviving Books, Silver & Furnishings* (Richmond, 1963), 19. Compare this to the twelve-to-seventeen foot average height of seventeenth-century New England Puritan meeting-houses.

13. For the importance of chancel screens to Anglican services and their use in Virginia, see Dorsey, *Early English Churches in America,* 16–17. For Jamestown's chancel see Strachey, *True Reportory,* 80. The evidence for a chancel screen at Jamestown is ambiguous, but they were common in later Virginia churches.

14. Upton, *Holy Things and Profane,* 74.

15. Upton argues that testers were not acoustically reflective but were merely for show. Many testers earlier in the century, however, even some that he presents, did serve an acoustic purpose, in part because of their placement. Upton's decorative testers became more prevalent as the decibel level of preaching styles diminished and pulpits were moved to visual rather than acoustic centers. The process was never complete, and acoustically effective pulpits were still prevalent later in the eighteenth century, particularly in more rural churches such as Christ Church in Broad Creek Hundred, near Laurel, Delaware.

16. Peter Benes, *New England Meeting House and Church, 1630–1850* (Boston, 1980), 35–43.

17. Nathaniel Bradstreet Shurtleff, ed., *Records of the Governor and Company of the Massachusetts Bay,* 5 vols. (Boston, 1853), 1:157, 181, 291; Lemuel Shattuck, *A History of the Town of Concord; Middlesex County, Massachusetts, from Its Earliest Settlement to 1832* (Boston, 1835), 7; Massachusetts Historical Society, *Winthrop Papers,* 5 vols. (Boston, 1929–47), 3: 181–82; and Donnelly, *New England Meeting Houses of the Seventeenth Century,* 17.

18. Felix Emil Held, ed., *Johann Valentin Andreae's Christianopolis,* cited in Donnelly, *New England Meeting Houses of the Seventeenth Century,* 33–35, 136 n. 41, 155.

19. Donnelly, *New England Meeting Houses of the Seventeenth Century,* 69–72.

20. Ibid., 130.

21. Robert J. Dinkin, "Seating the Meeting House in Early Massachusetts," *New England Quarterly* 43 (1970): 450–64, reprinted in *Material Life in America, 1600–1860,* ed. Robert Blair St. George (Boston, 1988), 407–18. Subsequent references cite the reprint.

22. Jane Neill Kamensky, *Governing the Tongue: The Politics of Speech in Early New England* (Oxford, 1998), 71–98, 169–79; and Carla Gardina Pestana, "The Quaker Executions as Myth and History," *Journal of American History* 80, no. 2 (1993): 441–69.

23. George W. Chase, *The History of Haverhill, Massachusetts, from Its First Settlement in 1640 to the Year 1860* (Haverhill, Mass., 1861), 265, cited in Dinkin, "Seating the Meeting House in Early Massachusetts," 413.

24. Dinkin, "Seating the Meeting House in Early Massachusetts," 413.

25. The material on octagonal and hexagonal acoustics is derived from a visit to a Birmingham meeting in Pennsylvania in February 1998. Also see T[homas] Chalkley Matlack papers, Haverford College Quaker collection, mss 1106, box 1 book 4: "Friend's meeting Houses in Pennsylvania: Columbia, Delaware, and Lycoming Counties," s.v. "Cheyney: the Shelter."

26. Donnelly, *New England Meeting Houses of the Seventeenth Century,* 79; John Thomson Faris, *Old Churches and Meeting Houses in and around Philadelphia* (Philadelphia, 1926), 79–80; "Bank Street Meeting House," Haverford College Quaker Collection, Philadelphia Meeting houses, 911 A-F Box 1; David A. Barton, *Discovering Chapels and Meeting Houses* (Aylesbury, Bucks, [1975]), 56; Hubert Martin Lidbetter, *The Friends Meeting House: An Historical Survey of the Places of Worship of the Society of Friends (Quakers), from the Days of Their Founder George Fox, in the Seventeenthth Century, to the Present Day* (York, [Eng.], [1961]), 21.

27. George Vaux, "The Great Meeting House," *Friend* 63 (1890): 148, Division of Com-

munity Service Programs, Pennsylvania Historical Survey, Works Project Admnistration, *Inventory of Church Archives, Society of Friends in Pennsylvania* (Philadelphia, 1941), 63.

28. Division of Community Service Programs, Pennsylvania Historical Survey, Works Project Admnistration, *Inventory of Church Archives, Society of Friends in Pennsylvania*, 63.

Chapter 4. On the Rant

1. I am marking here the resemblance of the notion of civil society as conceived by Jürgen Habermas and other modern political theorists to older ideas of "the civil" and "civilization," the latter attended by their well-worn binaries, the "other," the "wild," and the "savage." Habermas's defense of the modern liberal nation state as a sort of last best hope for rational and equitable government sneaks in the older idea of civilization in the guise of "civil society." The savage and the other are never named but always implied, waiting at the gates should modern liberal democracy fail. I am in some ways tracing the ancestry of this notion of civil society with the binary articulated. See the following by Jürgen Habermas: "The Public Sphere," *New German Critique* 1, no. 3 (1974): 49–55; *The Theory of Communicative Action* (Boston, 1984); *The Past as Future* (Cambridge, 1994); *The Structural Transformation of the Public Sphere: An Inquiry into a Category of Bourgeois Society,* Studies in Contemporary German Social Thought (Cambridge, Mass., 1991).

2. Robert Blair St. George has tracked New Englanders' concerns for the dangers of "heated" speech and their efforts to control them by assaying the disproportionately frequent (in comparison to today) litigation for speech crimes in the seventeenth century. Robert Blair St. George, " 'Heated' Speech and Literacy in Seventeenth-Century New England," in *Seventeenth-Century New England,* ed. David G. Allen and David D. Hall (Boston, 1985), 275–322.

3. John Smith, *"The Generall History of Virginia, New England, and the Summer Isles (1624),"* in *CWCJS,* 2: 129.

4. *OED Online,* s.v. "discourse," "discuss." For one early American example among many, see Increase Mather Diary, April 17, 1664, Mather Family Papers, box 3, AAS: "After sermon went to speake with Studley who was seduced by Quakers, she desiring to speake with me. Lord helped me in discourse and Prayer with her." Also see deposition of William Baker in 1651 in Middlesex County Court of Common Pleas Folio Collection at Columbia Point Archive, Boston.

5. Strachey, *True Reportory,* 40–45. See also Richard Allestree, *The Whole Duty of Man* (Williamsburg, 1746), 124.

6. Kathleen M. Brown, *Good Wives, Nasty Wenches, and Anxious Patriarchs: Gender, Race and Power in Colonial Virginia* (Chapel Hill, N.C., 1996), 90; Robert Barclay, *The Anarchy of the Ranters, and Other Libertines, the Hierarchy of the Romanists and Other Pretended Churches, Equally Refused, and Refuted, in a Two-fold Apology for the Church and People of God Called in Derision Quakers* (London, 1771), 35.

7. AAS, Mather Family Papers, Increase Mather, Box 3 Folder 2, Diary typescript, 1675–1676, ed. Michael G. Hall, Jan. 28, 1675/6; Cotton Mather, *Wonders of the Invisible World* (Boston, 1693), 68.

8. Robert Beverly, *The History and Present State of Virginia* (Chapel Hill, N.C., 1947).

9. Increase Mather, "The Autobiography of Increase Mather," *Proceedings of the American Antiquarian Society,* n.s. 71, no. 2 (1961): 271–360, ed. Michael G. Hall, 282; *OED Online,* s.v. "hum."

10. *OED Online,* s.v. "groaning"; Samuel Sewall, *Diary of Samuel Sewall, 1674–1729,* 2 vols. (New York, 1973), Feb. 16, 1677, Jan. 9, 1724; Brown, *Good Wives, Nasty Wenches, and Anxious Patriarchs,* 188, 424 n. 67.

11. Mather, "Autobiography," 291–94; Sermons on [of] Increase Mather, Mather Family Papers, AAS, Box 2, Folder 10, envelope c, 13; Cotton Mather, *Boanerges, a Short Essay to Preserve and Strengthen the Good Impressions Produced by Earthquakes on the Minds of People That Have Been Awakened with Them* (Boston, 1727), 23; John Danforth, *A Sermon Occasioned by the Late Great Earthquake and the Terror that Attended It* (Boston, 1728), 1–5, in second set of page numbers, after main text of pamphlet.

12. Increase Mather, *An Essay for the Recording of Illustrious Providences* (Boston, 1864), 170–71.

13. Ibid., 135–38, 171.

14. Ibid., 140–41, 170–71, 196.

15. Ibid., 153–54; Robert Calef, *More Wonders of the Invisible World* (London, 1700), 9.

16. John Gyles, *Memoirs of the Odd Adventures and Strange Deliverances, Etc., in the Captivity of John Gyles* (Boston, 1736).

17. J. William Frost, ed., *The Keithian Controversy in Early Pennsylvania* (Norwood, Penn., 1980); James Pinckney Bell, *Our Quaker Friends of Ye Olden Time* (Lynchburg, Virginia, 1905), 129, 148–50.

18. Kenneth L. Carroll, "Singing in the Spirit in Early Quakerism," *Quaker History* 73, no. 1 (1984): 1–13; Mather, *Illustrious Providences*, 341–57; George Keith, *The Presbyterian and Independent Visible Churches in New England and Elsewhere Brought to the Test* (London, 1691), 214–30.

19. Habermas, "Public Sphere"; Habermas, *Structural Transformation of the Public Sphere;* T. H. Breen, "Early America and the Public Sphere," paper delivered at the meeting of the Organization of American Historians, San Francisco, 1997.

20. Michael Warner, *The Letters of the Republic: Publication and the Public Sphere in Eighteenth-Century America* (Cambridge, 1990); Perry Miller, *Errand into the Wilderness* (Cambridge, Mass., 1956), 15.

21. Mather, *Illustrious Providences*, 341. For Ross's involvement in Harris's conversion see Keith, *Presbyterian and Independent Visible Churches*, 214; New Plymouth Colony et al., *Records of the Colony of New Plymouth, in New England* (Boston, 1855), 6: 113. Dunham was also spelled "Dunen" and "Denham," and he went by the alias Singletary, too.

22. Mather, *Illustrious Providences*, 135–37, 140–41, 154, 170–71, 341–42. For Harris as merchant, see Gale Ion Harris, "The Origins of Thomas Harris of East Hampton, Long Island and Killingworth, Connecticut," *New York Genealogical and Biographical Record* 128, no. 1 (1997): 12, 15–20. Thanks to Jerome E. Anderson of the New England Historical and Genealogical Society for bringing this article to my attention.

23. Mather, *Illustrious Providences*, 343–44 (italics added).

24. Ibid., 346–47; [Charles Leslie], *The Snake in the Grass* (London, 1689), 74–75; Kenneth Carroll, "Early Quakers and 'Going Naked as a Sign,'" *Quaker History* 67, no. 2 (1978): 69–87.

25. Mather, *Illustrious Providences*, 344; *OED Online*, s.v. "rail." *Singing* is of course transitive in one sense, but the possible direct objects are all songs or lyrics. In the seventeenth century, the verb "to sing" could take people as its direct object as well.

26. Mather, *Illustrious Providences*, 344–45.

27. *Records of the Colony of New Plymouth*, 6: 114.

28. Mather, *Illustrious Providences*, 347–56. David D. Hall, *Worlds of Wonder, Days of Judgement: Popular Religious Belief in Early New England* (Cambridge, Mass., 1990); David D. Hall, "The Uses of Literacy in New England, 1600–1850," in *Cultures of Print: Essays in the History of the Book*, ed. David D. Hall (Amherst, 1996), 97–150.

29. Jane Neill Kamensky, *Governing the Tongue: The Politics of Speech in Early New England* (Oxford, 1998), 71–98; Carla Gardina Pestana, "The Quaker Executions as Myth and History," *Journal of American History* 80, no. 2 (1993): 441–69.

30. Douglas Gwyn, *Apocalypse of the Word: The Life and Message of George Fox (1624–1691)*

(Richmond, Ind., 1986), 35; Richard Bauman, *Let Your Words Be Few: Symbolism of Speaking and Silence among Seventeenth-Century Quakers* (Cambridge, 1983), 64–67, 137–45; Baumann, 128–29, derives his views on routinization from Max Weber.

31. My thinking in this paragraph has been considerably sharpened by comments from Steve Marini of Wellesley College, though I remain responsible for any shortcomings.

32. For nonverbal vocalizations as a means of representing "otherness," see Erving Goffman, "Response Cries," *Language* 54, no. 4 (1978): 787–815, esp. 808–12.

33. Benedict Anderson, *Imagined Communities: Reflections of the Origin and Spread of Nationalism* (London, 1991); Warner, *Letters of the Republic*; Habermas, "Public Sphere."

34. Keith, *Presbyterian and Independent Visible Churches in New England and Elsewhere Brought to the Test*, 217.

35. Ibid., 215–17. For the credibility of Mather's sources, see Mather, *Illustrious Providences*, 344.

36. Keith, *Presbyterian and Independent Visible Churches in New England and Elsewhere Brought to the Test*, 214; William Edmundson, *Journal of the Life, Travels, Sufferings, and Labour of Love in Work of the Ministry, of That Worthy Elder, and Faithful Servant of Jesus Christ, William Edmundson* (Dublin, 1715), 91–92; James Dickinson, *Journal of the Life, Travels, and Labour of Love in the Work of the Ministry, of That Worthy Elder, and Faithful Servant of Jesus Christ, James Dickinson* (London, 1745); Thomas Story, *Journal of the Life of Thomas Story* (Newcastle upon Tyne, 1747), 192–193, 220; George Fox, *Journal or Historical Account of the Life Travels, Sufferings, Christian Experiences, and Labour of Love in the Work of the Ministry of that Ancient, Eminent and Faithful Servant of Jesus Christ, George Fox* (New York, 1975), 127–28; David S. Lovejoy, *Religious Enthusiasm in the New World: Heresy to Revolution* (Cambridge, Mass., 1985), 140–43.

37. William Penn, *Brief Examination and State of Liberty Spiritual, Both with Respect to Persons in Private Capacity, and in Their Church Society and Communion* (London, 1771), 106; Bauman, *Let Your Words Be Few*.

38. Penn, *Brief Examination*, 99. On tapping coins to hear their authenticity see Mark M. Smith, *Listening to Nineteenth-Century America* (Chapel Hill, N.C., 2001), 275 n. 21.

39. Barclay, *Anarchy of the Ranters*, 20, 35.

40. Penn, *Brief Examination*, 107. For the idea of social death as a very different context, see Orlando Patterson, *Slavery and Social Death: A Comparative Study* (Cambridge, 1982).

41. Lovejoy, *Religious Enthusiasm in the New World*, 99; [Leslie], *Snake in the Grass*, 74; Story, *Journal*, 192–93; Keith, *Presbyterian and Independent Visible Churches*, 216. One exception to non-self-identification was John Pearce of Elizabethtown, in what was then East Jersey, who wavered between Quakerism and a group he identified as Ranters, ultimately settling on affiliation with the latter. See Brendan McConville, "Confessions of an American Ranter," *Pennsylvania History* 62, no. 2 (1995): 238–48.

42. [Leslie], *Snake in the Grass*.

43. *OED Online*, s.v. "rant" and its derivatives. Robert Blair St. George, "'Heated' Speech," 275–322.

44. J. C. Davis, *Fear, Myth, and History: The Ranters and the Historians* (Cambridge, 1986); Christopher Hill, "Abolishing the Ranters," in *A Nation of Change and Novelty: Radical Politics, Religion, and Literature in Seventeenth Century England* (London, 1990), 152–94; David Underdown, *Revel, Riot, and Rebellion: Popular Politics and Culture in England 1603–1660* (Oxford, 1985), 249–50; Jerome Friedman, *Blasphemy, Immorality, and Anarchy: The Ranters and the English Revolution* (Athens, Ohio, 1987); Christopher Hill, *The World Turned Upside Down: Radical Ideas during the English Revolution* (New York, 1972); Arthur Leslie Morton, *The World of the Ranters: Religious Radicalism in the English Revolution* (London, 1970).

45. Story, *Journal*, 220. Lacking useful first-person accounts from anyone self-identifying as a "Ranter" or "Singing Quaker," the description of ranting here is a composite drawn by comparing Quaker, Puritan, and Anglican descriptions. See Story, *Journal*, 192–93; Sarah Kem-

ble Knight, "The Journal of Madam Knight," in *The Puritans: A Sourcebook of Their Writings*, 2 vols., ed. Perry Miller and Thomas H. Johnson (New York, 1963), 2: 445; [Leslie,] *Snake in the Grass*, 74–75; Mather, *Illustrious Providences*, 344–56; Keith, *Presbyterian and Independent Visible Churches*, 214–30; Edmundson, *Journal*, 91–92; Dickinson, *Journal*; George Fox, *Journal*, Chapter 18. Also see Lovejoy, *Religious Enthusiasm*, 140–43, and Carroll, "Singing in the Spirit in Early Quakerism." For a tape of present-day Quakers employing what may be a distant relative of this singing style, see "Intoning by Quaker Preachers" [audio tape], recorded by Thomas E. Jones and Esther Balderston at Earlham College (1974), Quaker Collection, Haverford College.

46. Phyllis Mack, *Visionary Women: Ecstatic Prophecy in Seventeenth-Century England* (Berkeley, 1992), 286, 66–67; Lovejoy, *Religious Enthusiasm in the New World*, 143; David D. Hall, ed., *The Antinomian Controversy, 1636–1638: A Documentary History* (Middletown, Conn., 1968), 268–70, 274.

47. Peter Ross, *History of Long Island from Its Earliest Settlement to Present Times*, 3 vols. (New York, 1905), 1: 165–75; Benjamin F. Thompson, *History of Long Island from Its Discovery and Settlement to the Present Time*, 4 vols. (New York, 1918); Arthur J. Worrall, *Quakers in the Colonial Northeast* (Hanover, N.H., 1980), 4–110.

48. *DRCHNY*, 14: 563–64.

49. *DRCHNY*, 3:415, *DRCHNY*, 14: 763.

50. Keith, *Presbyterian and Independent Visible Churches*, 195–203; Cotton Mather, *Magnalia Christi Americana, Books I and II* (1702; Cambridge, Mass., 1977).

51. Keith, *Presbyterian and Independent Visible Churches*, 217.

52. The account of the Keithian controversy below is drawn from Lovejoy, *Religious Enthusiasm in the New World*, 143–48; Ethyn Kirby, *George Keith (1638–1716)* (New York, 1942), Chapter 4; Rufus Matthew Jones, Isaac Sharpless, and Amelia Mott Gummere, *The Quakers in the American Colonies* (London, 1911); Charles P. Keith, *Chronicles of Pennsylvania from the English Revolution to the Peace of Aix-la-Chapelle 1688–1748*, 2 vols. (Philadelphia, 1917), 2: Chapter 8. The primary documents are collected in Frost, *Keithian Controversy*.

53. Philadelphia Society of Friends, "Minutes of the Yearly Meeting and General Spring Meeting of Ministers from the Fifth Month 1686/7 to Seventh Month 1710, and for Ministers and Elders from That Time to Seventh Month 1734," Haverford College Quaker Collection, MSS B2.1 (Philadelphia, 1686–1734), 14–31.

54. Quoted in Thompson, *History of Long Island*, 144.

55. Knight, "Journal," 445.

56. On "unsaying," see Kamensky, *Governing the Tongue*, 127–49.

57. George Keith, *The Quakers Proved Apostats and Heathens* (London, 1700), 2–3.

Chapter 5. The Howling Wilderness

1. Ulali home page, www.ulali.com.

2. Perry Miller, *Errand into the Wilderness* (Cambridge, Mass., 1956), 1–2.

3. On landscapes as human constructions, see Simon Schama, *Landscape and Memory* (New York, 1995), 7, 144. My reading of wilderness leans heavily on Roderick Nash, *Wilderness and the American Mind* (New Haven, 1982), 1–3, 8, 10–17, 24–38. For the concept of wildness as "other," see Richard Bernheimer, *Wild Men in the Middle Ages: A Study in Art, Sentiment, and Demonology* (Cambridge, Mass., 1952), and Michael Taussig, *Shamanism, Colonialism, and the Wild Man: A Study in Terror and Healing* (Chicago, 1987). On the frontier, see Frederick Jackson Turner, *The Significance of the Frontier in American History* (Madison, 1894), and Ray Allen Billington, *America's Frontier Heritage* (New York, 1966).

4. Edward Johnson, *The Wonder-Working Providence of Sion's Savior in New England* (New

York, 1910), 115; Thomas Hooker, *The Application of Redemption by the Effectual Work of the Word, and Spirit of Christ, for the Bringing Home of Lost Sinners to God* (London, 1656), 5–13; Miller, *Errand into the Wilderness*, 15. Karen Ordahl Kupperman notes that the term "howling wilderness" was used nostalgically by the second generation and very little by the first generation. See Karen Ordahl Kupperman, "Climate and Mastery of the Wilderness in Seventeenth-Century New England," in *Seventeenth-Century New England*, ed. David D. Hall and David Grayson Allen (Boston, 1984), 24–25, 25 n. 5.

5. Wigglesworth, "God's Controversy with New England," *Proceedings of the Massachusetts Historical Society* 12 (1871–73): 83–84.

6. Mary Rowlandson, *Narrative of the Captivity and Restoration of Mrs. Mary Rowlandson* (n.p., 1997), "The Sixth Remove," http://ibiblio.org/gutenberg/etext97/crmmr10.txt.

7. Rowlandson, *Narrative of the Captivity and Restoration*, Introduction, "The First Remove," "The Third Remove," and "The Thirteenth Remove," http://ibiblio.org/gutenberg/etext97/crmmr10.txt; and AAS, "Correspondence of Increase Mather," Box 2, Folder 9, 17. William Strachey also noted that at deaths, native Chesapeake women painted their faces black to mourn and go "lamenting by turnes with such yelling and howling, as may express their great passions." William Strachey, *The Historie of Travell into Virginia Britania (1612)* (London, 1953), 95.

8. *CWCJS*, 1: 11–12, 59; Strachey, *Historie of Travell into Virginia*, 71, 87, 96–97; John Smith, "*The Proceedings of the English Colonie in Virginia, [1606–1612] (1612)*," in *CWCJS*, 1: 236; John Smith, "*The Generall History of Virginia, New England, and the Summer Isles (1624)*," in *CWCJS*, 2: 183.

9. Smith, *True Relation*, in *CWCJS*, 1: 45; Strachey, *Historie of Travell into Virginia Britania*, 59; Briton Hammon, *A Narrative of the Uncommon Sufferings, and Surprizing Deliverance of Briton Hammon, a Negro Man* (Boston, 1760), 6; *OED Online*, s.v. "holler," "hollo," "halloo," "hello," "hillo," "hullo," "hollo," "hallow," "holla," "howl."

10. Carl Bridenbaugh, ed., *Gentleman's Progress: The Itinerarium of Dr. Alexander Hamilton, 1744* (Chapel Hill, N.C., 1948), 70–71. Thanks to John Brooke for bringing this incident to my attention.

11. Henry David Thoreau, *A Week on the Concord and Merrimack Rivers; Walden, Or, Life in the Woods; The Maine Woods; Cape Cod* (New York, 1985), 761.

12. Eliot only modifies the names of the alphabet's letters slightly in order to simplify them and make them a better fit for the Indian sounds they described. For example, he eliminates the letter "c" in favor of "k" and "s," retaining "c" only for its use in words such as "chew," and he distinguishes the vowels "i" and "u" from the consonants "j" and "v," foreshadowing current practices. The only other changes he makes are adding marks to indicate the proper accentuation of syllables and adding the diacritic "^" to indicate nasalization of vowels. A non-alphabetic symbol, "8," is introduced, but it simply replaces "oo" as a spelling rather than as a pronunciation, as he gives as examples both "moody" and "brook" (though perhaps this could instead be taken as evidence that both were pronounced with the same vowel). The point is that this orthographic innovation was not a phonetic innovation. John Eliot, *The Indian Grammar Begun: or an Essay to Bring the Indian Language in to Rules, for the Help of Such as Desire to Learn the Same, for the Furtherance of the Gospel among Them* (Cambridge, Mass., 1666), 11–12.

13. Ibid., 4, 22.

14. Smith, *Map of Virginia*, in *CWCJS*, 1: 149; Smith, *Proceedings*, in CWCJS, 1: 227; Smith, *Generall Historie*, in *CWCJS*, 2: 166; George Percy, "Observations Gathered out of a Discourse of the Plantation of the Southerne Colonie in Virginia by the English, 1606," in *The Jamestown Voyages Under the First Charter, 1606–1609*, ed. Philip L. Barbour (London, 1969), 2: 136, 143.

15. Eliot, *Indian Grammar Begun*, 6, 23, 25; Henry Whitfeld, *The Light Appearing More and*

More Towards the Perfect Day (London, 1651), 2. Thanks to David Murray for pointing out the Eliot verse.

16. Bernardino de Sahagún, *Florentine Codex*, Book 6, Chapter 22, quoted in Walter D. Mignolo, *The Darker Side of the Renaissance: Literacy, Territoriality, and Colonization* (Ann Arbor, 1995), 117–18.

17. Père Chrestien LeClerq, *New Relation of Gaspesia with the Customs and Religion of the Gaspesian Indians*, ed. and trans. William F. Ganong (New York, 1968), 311, cited in Gordon M. Sayre, *Les Sauvages Américains: Representations of First Nations people in French and English Colonial Literature* (Chapel Hill, N.C., 1997), 274.

18. *H5N*, 3–6.

19. *H5N*, 6.

20. Smith, *Map of Virginia*, in *CWCJS*, 1: 167; Smith, *Generall Historie*, in *CWCJS*, 2: 120, 245; Smith, *True Relation*, in *CWCJS*, 1: 57, 69; Mather, *Illustrious Providences*, 39; *H5N*, 23; John Gyles, *Memoirs of the Odd Adventures and Strange Deliverances, Etc., in the Captivity of John Gyles* (Boston, 1736), 2.

21. *H5N*, 6; Smith, *Map of Virginia*, in *CWCJS*, 1: 167; Smith, *Generall Historie*, in *CWCJS*, 2: 120; Strachey, *Historie of Travell into Virginia*, 85; David D. Field, *A Statistical Account of the County of Middlesex, in Connecticut* (Middletown, Conn., 1819), 71–73.

22. [William Smith], *Some Account of the North American Indians; Their Genius, Characters, Customs, and Dispositions, Towards the French and English Nations* (London, 1754), 26–27.

23. Gregory Evans Dowd, *A Spirited Resistance: The North American Indian Struggle for Unity, 1745–1815* (Baltimore, 1993), 49, xv.

24. *H5N*, 7.

25. *H5N*, 7–8.

26. *H5N*, 9–10.

27. The choice of execution or torture generally lay with the matriarch of the deceased's family. Adoptions helped replace wartime population losses. This was usually not a crucial function, though, as in any population it is ultimately the number of women rather than the number of men that secures following generations. Thus execution was often the route taken with no long-term consequences to either the captives' or captors' populations. The system of adoption or execution worked under normal circumstances but became dysfunctional after European contact caused dramatic rises in deaths both by war and disease. Wars of mourning and replacement became more frequent, and their very frequency past a certain limit caused even more wars with escalating losses. At the same time, disease decimated men and women alike, causing a demographic disaster of epic proportions on the continent. Daniel K. Richter, *The Ordeal of the Longhouse: The Peoples of the Iroquois League in the Era of European Colonization* (Chapel Hill, N.C., 1992), 32–37, 148–49. Notice that this system of adoption contradicts the blood quantum standard of Indian "tribal" ethnicity in use by the United States government today for federal recognition of tribal status. Native American notions of ethnicity operate completely in the cultural realm rather than in the biological.

28. [Smith], *Some Account of the North American Indians*, 24–25.

29. Jonathan Carver, *Travels through the Interior Parts of North-America, in the Years 1766, 1777, and 1778* (London, 1778), 334.

30. *H5N*, 62, 122–23.

31. *JR* 42: 191, 193.

32. Richter, *Ordeal of Longhouse*, 66–67; James Axtell, "The White Indians," in James Axtell, *The Invasion Within: The Contest of Cultures in Colonial North America* (Oxford, 1985), 210, 375 n. 34; Bruce G. Trigger, *The Huron: Farmers of the North* (Fort Worth, 1990), 58.

33. Strachey, *Historie of Travell into Virginia*, 85–86; Field, *Statistical Account of the County of Middlesex*, 72.

34. *H5N*, 9–10.

35. *H5N*, 10–12.

36. *H5N*, 13.

37. *H5N*, 39; [Smith], *Some Account of the North American Indians*, 25.

38. Carver, *Travels through the Interior Parts of North-America*, 260–61; Richter, *Ordeal of Longhouse*, 46–47; Virginia, *Treaty Held with the Catawba and Cherokee Indians* (Williamsburg, 1756), 2; Smith, *Generall History*, in *CWCJS*, 2: 246–47.

39. [Conrad Weiser], *The Treaty Held with the Indians of the Six Nations, at Lancaster, Pennsylvania, in June, 1744* (Williamsburg, 1744), ix. The Tuscaroras were admitted to the Iroquois Confederacy in the 1720s, making the Five Nations six.

40. Pastor Waldeck, "Diary, 1776–1781," in Hessian Manuscripts number 2, typescript, trans. E. Bruce Burgoyne, New York Public Library, 203b; [Smith] *Some Account of the North American Indians*, 27; Carver, *Travels through the Interior Parts of North-America*, 90.

41. *H5N*, 44.

42. *JR* 42: 77, 79.

43. For a close examination of indigenous "books" and the limits of Eurocentric analogies in early colonial Mexico, see Mignolo, *Darker Side of the Renaissance*, 10, 69–81, 188–89. For a discussion of metaphor that is akin to my critique of similes, see Murray, *Indian Giving*, 10–11.

44. Peter Wraxall, "Proceedings of the Congress at Albany, 1754," in codex=Eng 36, MS, JCB, 29–30. For a sampling of Wampum and pelts used to punctuate seventeenth- and eighteenth-century speeches of English, Dutch, Iroquois, French, and Esopus treaty makers, see *DRCHNY*, 13:72, 104–5, 107–10, 126; *H5N*, 30–31, 37, 51, 57; *JR* 42:37, 47, 99, 101, 167, 187, 189; [Smith], *Some Account of the North American Indians*, 32–33; Wraxall, "Proceedings of the Congress at Albany, 1754," 8–9, 20–24, 27–30, 34, 35, 38, 40–42, 69–72, 78. I am disagreeing with Richter's assessment of the Iroquois League as a nonstate. A state based on persuasion rather than coercion is sufficiently different from European definitions of state as to perhaps warrant Richter's claim, but they were called nations, treated as nations, and to the extent they could persuade their populace, acted as nations in a sense close enough for Europeans of the time. By the mid-seventeenth century, the parallel Iroquois confederacy, which served to deal with Europeans on a political basis, had emerged, making Iroquois more statelike on European terms. Richter, *Ordeal of the Longhouse*, 40–41, 169–71.

45. Richter, *Ordeal of the Longhouse*, 47; Wraxall, "Proceedings of the Congress at Albany, 1754," MS, codex=Eng 36 manuscript, JCB, 38. For an excellent contribution to and summary of the cultural meanings of wampum, see Murray, *Indian Giving*, 116–40.

46. Jonathan Carver, "Journals of the Travels of Jonathan Carver," in *The Journals of Jonathan Carver and Related Documents, 1766–1770*, ed. John Parker (St. Paul, 1976), 143–44; Joseph-François Lafitau, *Moeurs Des Sauvages Ameriquains, Comparées Aux Moeurs Des Premiers Temps* (Paris, 1724), 328, cited in Sayre, *Les Sauvages Américains*, 190–91.

47. *DRCHNY* 14:369 (emphasis added). For a critical discussion of the concept of the gift, as well as an overview of the vast anthropological literature on the subject, see Murray, *Indian Giving*, 20–39.

48. *H5N*, 23, 36.

49. *H5N*, 51, 52, 55 (emphasis in original). All the punctuating quotations are set off in italics in the print version.

50. *Maryland Gazette*, March 21 and March 28, 1754, reprinted in full at http://earlyamerica.com/earlyamerica/milestones/journal/journaltext.html.

51. *H5N*, 27.

52. James H. Merrell, *The Indian's New World: Catawbas and Their Neighbors from European Contact through the Era of Removal* (New York, 1989), 160–164; Francis Jennings, "Iroquois Alliances in America History," in *The History and Culture of Iroquois Diplomacy: An Interdisciplinary*

Guide to the Treaties of the Six Nations and Their League, ed. Francis Jennings et al. (Syracuse, 1985), 50–52.

53. Virginia, *Treaty Held with the Catwba and Cherokee Indians,* vi.
54. Ibid., iv, vii, xi–xii.
55. Ibid., 1–6.
56. Ibid., xi–xii.
57. *H5N,* 159; Merrell, *Indian's New World,* 148–149; William N. Fenton, "Structure, Continuity, and Change in the Process of Iroquois Treaty Making," in *History and Culture of Iroquois Diplomacy,* 115.
58. [Weiser], *Treaty Held with the Indians of the Six Nations,* viii–ix; *H5N,* 39; *DRCHSNY,* 13:102–103; Waldeck, "Diary," 42b.
59. Wraxall, "Proceedings of the Congress at Albany," 68.
60. G. Peter Jemison, Anna M. Schein, and John Mohawk, eds., *Treaty of Canandaigua 1794: 200 Years of Treaty Relations between the Iroquois Confederacy and the United States* (Santa Fe, 2000); George C. Shattuck, *The Oneida Land Claims: A Legal History* (Syracuse, 1991).

Conclusion: Worlds Chanted into Being

1. Robert Blair St. George, "'Heated' Speech and Literacy in Seventeenth-Century New England," in *Seventeenth-Century New England,* ed. David G. Allen and David D. Hall (Boston, 1985), 275–322; Mary Beth Norton, *Founding Mothers & Fathers: Gendered Power and the Forming of American Society* (New York, 1996); Kathleen M. Brown, *Good Wives, Nasty Wenches, and Anxious Patriarchs: Gender, Race and Power in Colonial Virginia* (Chapel Hill, N.C., 1996); and Jane Neill Kamensky, *Governing the Tongue: The Politics of Speech in Early New England* (Oxford, 1998).
2. For word magic, see Keith Thomas, *Religion and the Decline of Magic: Studies in Popular Beliefs in Sixteenth and Seventeenth-Century England* (New York, 1971).
3. David Cressy, *Bonfires and Bells: National Memory and the Protestant Calendar in Elizabethan and Stuart England* (Berkeley, 1989); David Waldstreicher, *In the Midst of Perpetual Fetes: The Making of American Nationalism, 1776–1820* (Chapel Hill, N.C., 1997); and Len Travers, *Celebrating the Fourth: Independence Day and the Rites of Nationalism in the Early Republic* (Amherst, 1997).
4. Julius Scott, "The Common Wind: Currents of Afro-American Communication in the Era of the Haitian Revolution" (Ph.D. diss., Duke University, 1986); T. H. Breen, "'Baubles of Britain': The American and Consumer Revolutions of the Eighteenth Century," *Past and Present* 119 (1988): 73–104; Peter Linebaugh and Marcus Buford Rediker, *The Many-Headed Hydra: Sailors, Slaves, Commoners, and the Hidden History of the Revolutionary Atlantic* (Boston, 2000).
5. Richard L. Merritt, *Symbols of American Community, 1735–1775* (New Haven, 1966); Richard D. Brown, *Knowledge Is Power: The Diffusion of Information in Early America, 1700–1865* (New York, 1989).
6. David W. Conroy, *In Public Houses: Drink and the Revolution in Authority in Colonial Massachusetts* (Chapel Hill, N.C., 1995); Peter Thompson, *Rum Punch & Revolution: Taverngoing & Public Life in Eighteenth Century Philadelphia* (Philadelphia, 1999).
7. Benedict Anderson, *Imagined Communities: Reflections of the Origin and Spread of Nationalism* (London, 1991); Michael Warner, *The Letters of the Republic: Publication and the Public Sphere in Eighteenth-Century America* (Cambridge, Mass., 1990), 13–14.
8. Bernard Bailyn, *The Ideological Origins of the American Revolution* (Cambridge, 1968); Anderson, *Imagined Communities.*
9. Harry Stout makes the same point, backing his argument with evidence from eigh-

teenth-century sermons and oral culture. The present work underscores some of Stout's findings by providing a reason why oral culture was so crucial to the nation-building process. Rather than finding a complete, non-elite ideology there, I am pointing toward some of the same connections with print and the public sphere Frank Lambert and David Waldstreicher have described, and advocating that their findings and methods be brought together with those of Linebaugh and Rediker and Scott. Harry S. Stout, "Religion, Communications, and the Ideological Origins of the American Revolution," *William and Mary Quarterly* 34, no. 4 (1977): 519–41; Frank Lambert, *"Pedlar in Divinity": George Whitefield and the Transatlantic Revivals, 1737–1770* (Princeton, 1994); Waldstreicher, *In the Midst of Perpetual Fetes*; Scott, "The Common Wind"; Linebaugh and Rediker, *Many-Headed Hydra*.

10. T. H. Breen, "Ideology and Nationalism on the Eve of the American Revolution: Revisions Once More in Need of Revising," *Journal of American History* 84, no. 1 (1997): 13–39; Carroll Smith-Rosenberg, "Discovering the Subject of the 'Great Constitutional Discussion,' 1786–1789," *Journal of American History* 79 (1992): 841–73; Travers, *Celebrating the Fourth*; Waldstreicher, *In the Midst of Perpetual Fetes*.

11. Jürgen Habermas, *The Structural Transformation of the Public Sphere: An Inquiry into a Category of Bourgeois Society* (Cambridge, Mass., 1991). The most thorough employment of Habermas's model to the American situation is Warner, *Letters of the Republic*.

12. Miriam Hansen, *Babel and Babylon: Spectatorship in American Silent Film* (Cambridge, Mass., 1991).

13. *OED Online*, s.v. "text."

14. Both verbs carried other meanings at that time. "Discourse" comes from the Latin words meaning "to and fro" and "to run" (the same root as "courier") and literally meant to "run back and forth" as well as its primary meaning of discussing or conversing. "Discuss" comes from the Latin word for "to shake asunder," which in later Latin came to mean "to verbally examine something." Although the words have different etymologies, they converged in seventeenth-century English. With the onset of [r] loss following a vowel in the emerging Standard English of the day, the two words converged in meaning and usage into the word we now know as discuss, with "discourse" seldom used as a verb at all anymore. *OED Online*, s.v. "discourse," "discusss." On [r] loss, see Richard Cullen Rath, "'What Meanes Hee May for to Gett Her Over': The Transference of Language and Culture from Old to New England in the Seventeenth Century" (honors thesis, Millersville University, 1991).

15. Marshall McLuhan, *The Gutenberg Galaxy: The Making of Typographic Man* (New York, 1962), 249.

Index

Within this index numbers followed by fig. *refer to illustrations and numbers followed by a* t *refer to tables.*

acoustic spaces, public, 180. *See also* public spheres

acoustics: contemporary, 98; of European churches, 98–103; invisible worlds and, 111–13; noise and, 116; of North American churches, 103–7, 208nn6–7, 209nn12–13, 209n15; meetinghouses, 3, 103, 107–19; seating patterns and, 103–5, 112–13, 118–19; theaters, 102*fig.*, 103, 109–110; silence and, 116; social order and, 104–5, 110–13, 116, 118–19; sounding boards, 98–100, 102–4, 106–7, 114–15, 118–19, 208n6, 209n15; vision and, 98–99, 104–5, 113, 208n7

acts of identity, 28–39, 57–61, 196n41

Adis, Henry, 20

African Americans: call and response, 8–9, 160; communication networks and, 177, 179; creolization/pidginization, 70–72, 96; cross-cultural communication and, 46; dance, 91–94; ethnicities, 83–85; instrumental sounds and, 46; Jamaica and, 8–9; martial arts, 86–89; military music, 94–96; orality/literacy, 2–3; possession and, 126; resistance, 85–89; slavery and, 175; thunder and, 39–41; the wilderness and, 149. *See also* Africans; enslaved Africans

African diaspora, 83–85

Africans: court music, 79–80; cross-cultural communications, 88–89; dance, 91–94; drums and drumming, 78–80; ethnicities, 83–85, 88; stringed instruments, 80–83. *See also* African Americans; enslaved Africans

agency, 16–21

Akan, 72, 75–76, 202n51

alarms, 57–61

Algonquians, 28, 30, 152–55

Allestree, Richard, 6

Allouez, Claude, 193n2

American Revolution: military music, 94–96, 179

amplification: of bells, 43–44, 46, 56–57; in churches, 99–100, 103, 115, 207n5; drums and, 5; earshot range and, 50–51, 56–57

Anang, 8

Anarchy of the Ranters, 135, 139

anarchy, 135–36, 138

Anderson, Benedict, 143, 178–79

André, Louis, 34–35

Andreae, Johann Valentin, 109–10

Anglo-Americans, 56–57, 91–92, 98–103, 123, 126, 175, 204n69

Angola, 72–77, 81, 86–87

animekeek, 31

animism, 29–30

Anokye, 40–41

Antinomian Controversy, 5, 132

antiphony, 8–9, 159–61

Arch Street Meetinghouse, 118*fig.*

Archer, Gabriel, 15

Arnold, Samuel, 1, 10

Ashanti, 40–41

assents, 8–9, 159–61

audience, 177, 180. *See also* public sphere

Bach, Johann Sebastian, 102

Bache, Humphrey, 19